Lecture Notes in Statistics

Edited by D. Brillinger, S. Fienberg, J. Gani,
J. Hartigan, J. Kiefer, and K. Krickeberg

9

Bent Jørgensen

Statistical Properties of the Generalized Inverse Gaussian Distribution

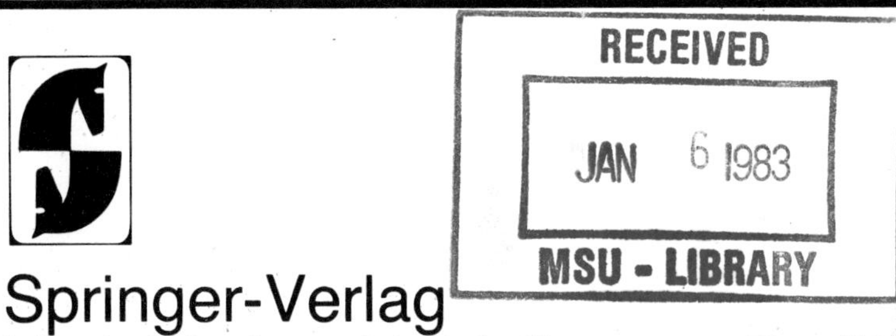

Springer-Verlag
New York Heidelberg Berlin

Bent Jørgensen
Department of Mathematics
Odense University
Campusvej 55
DL-5230 Odense M
Denmark

AMS Classification: 62E99

Library of Congress Cataloging in Publication Data

Jørgensen, Bent.
Statistical properties of the generalized inverse Gaussian distribution.

(Lecture notes in statistics; 9)
Originally presented as the author's thesis (M.Sc.)—Aarhus University.
Bibliography: p.
Includes index.
1. Gaussian distributtion. I. Title. II. Series: Lecture notes in statistics (Springer-Verlag); v. 9.
QA276.7.J67 1982 519.5'3 81-18453
AACR2

Printed in the United States of America.

9 8 7 6 5 4 3 2 1

ISBN 0-387-90665-7 Springer-Verlag New York Heidelberg Berlin
ISBN 3-540-90665-7 Springer-Verlag Berlin Heidelberg New York

Preface

In 1978 the idea of studying the generalized inverse Gaussian distribution was proposed to me by Professor Ole Barndorff-Nielsen, who had come across the distribution in the study of the socalled hyperbolic distributions where it emerged in connection with the representation of the hyperbolic distributions as mixtures of normal distributions. The statistical properties of the generalized inverse Gaussian distribution were at that time virtually undeveloped, but it turned out that the distribution has some nice properties, and models many sets of data satisfactorily. This work contains an account of the statistical properties of the distribution as far as they are developed at present.

The work was done at the Department of Theoretical Statistics, Aarhus University, mostly in 1979, and was partial fulfilment towards my M.Sc. degree. I wish to convey my warm thanks to Ole Barndorff-Nielsen and Preben Blæsild for their advice and for comments on earlier versions of the manuscript and to Jette Hamborg for her skilful typing.

Bent Jørgensen

Contents

1. Introduction

The subject of the present study is the generalized inverse Gaussian distribution whose probability density function is given by

$$\frac{(\psi/\chi)^{\frac{\lambda}{2}}}{2K_\lambda(\sqrt{\chi\psi})} x^{\lambda-1} e^{-\frac{1}{2}(\chi x^{-1} + \psi x)} \qquad (x>0), \qquad (1.1)$$

where K_λ is the modified Bessel function of the third kind and with index λ. Special cases of (1.1) are the gamma distribution ($\chi = 0$, $\lambda > 0$), the distribution of a reciprocal gamma variate ($\psi = 0$, $\lambda < 0$) (in the following denoted the reciprocal gamma distribution), the inverse Gaussian distribution ($\lambda = -\frac{1}{2}$) and the distribution of a reciprocal inverse Gaussian variate ($\lambda = \frac{1}{2}$). Other important cases are $\lambda = 0$ (the hyperbola distribution) and $\lambda = 1$.

Even though the generalized inverse Gaussian distribution was proposed more than 35 years ago by Good (1953) little is known about its statistical properties. It is the purpose of the present study to derive such properties, and our principal results concern the estimation and inference for the distribution.

The generalized inverse Gaussian distribution has been used by Sichel (1974, 1975) to construct mixtures of Poisson distributions, and by Barndorff-Nielsen (1977, 1978b), who has obtained the generalized hyperbolic distribution as a mixture of normal distributions, the mixing distribution in both cases being (1.1). In the latter case the mixture obtained for $\lambda = 1$ is the hyperbolic distribution which is characterized by having a hyperbolic log-density.

For some other uses of the generalized inverse Gaussian distribution see Wise (1975) and Marcus (1975). Blæsild (1978) has computed moments and cumulants and also considered the shape of the density.

Some of the probabilistic properties of the distribution have been investigated by Barndorff-Nielsen and Halgreen (1977), who have shown that (1.1) is infinitely divisible, and by Halgreen (1979), who has shown that (1.1) is self-decomposable.

The hyperbola distribution $(\lambda = 0)$ has been studied by Barndorff-Nielsen (1978b) and by Rukhin (1974). Both these authors observed that the hyperbola distribution yields a close analogy, for observations on a hyperbola, to the von Mises distribution on the circle, and we shall see that to some extent the analogy carries over to general λ. (For the von Mises distribution refer to Mardia (1972).)

Although the generalized inverse Gaussian distribution is mainly intended as a distribution for positive variates it proves useful to have the analogy to the von Mises distribution in mind, and of course one can identify a positive number x with the point (x,x^{-1}) on the hyperbola. However, for measurements of for example speed it is often a matter of taste whether one would record x or the reciprocal of x, a fact which indicates a certain relation to the hyperbola for such observations.

We shall see that the generalized inverse Gaussian distribution exhibits a certain analogy to the normal distribution, in much the same way as does the von Mises distribution, but it will be obvious from our results that the very close analogy between the normal and the inverse Gaussian distributions does not carry over to the generalized inverse Gaussian distribution. The special properties of the inverse Gaussian distribution are not covered systematically here; instead we refer to a recent review given by Chhikara and Folks (1978). In order to complete the very extensive list of references given in that paper we have included in our references a number of titles concerning the inverse Gaussian distribution.

Now, let us outline the contents of the chapters to follow.

Chapter 2 contains a review of the basic properties of the distribution. In particular we examine the shape of the density and give expressions for moments and cumulants, extending the results of Blæsild (1978).

Chapter 3 concerns a number of distributions related to the generalized inverse Gaussian. In particular we consider some normal approximations, the ratio and the product of two independent generalized inverse Gaussian variates, and powers and logarithms of generalized inverse Gaussian variates. We also define a generalized inverse Gaussian Markov process, and we consider the relation between the generalized hyperbolic and generalized inverse Gaussian distributions.

Chapter 4 concerns maximum likelihood estimation. The estimates are not obtainable in closed form, but a simple procedure for numerical calculation of the estimates is given and some approximate results are considered. We also consider the asymptotic distribution of the estimates.

In chapter 5 we consider the structure of the inference for

the distribution. We shall mainly be concerned with the inference about the parameters λ and $\omega = \sqrt{\chi\psi}$, and we also consider a one-way analysis of variance model and a regression model.

In chapter 6 we consider the generalized inverse Gaussian distribution as a lifetime model and we examine the shape of the hazard function for the distribution.

Finally, in chapter 7 we analyze some sets of data (picked from the literature) using the generalized inverse Gaussian distribution.

2. Basic properties

In this chapter we give some basic definitions and results which are used throughout the work.

The domain of variation for the parameters in (1.1) is given by

$$\lambda \in \mathbb{R}, \qquad (\chi,\psi) \in \Theta_\lambda,$$

where

$$\Theta_\lambda = \begin{cases} \{(\chi,\psi): \ \chi \geq 0, \ \psi > 0\} & \text{if } \lambda > 0 \\ \{(\chi,\psi): \ \chi > 0, \ \psi > 0\} & \text{if } \lambda = 0 \\ \{(\chi,\psi): \ \chi > 0, \ \psi \geq 0\} & \text{if } \lambda < 0. \end{cases} \tag{2.1}$$

In the cases $\chi = 0$ and $\psi = 0$ the norming constant in (1.1) is found using (A.1) and the asymptotic relation (A.7) from the Appendix, where a number of important results concerning the Bessel functions have been collected.

We use the symbol $N^{-}(\lambda,\chi,\psi)$ for the distribution (1.1) and we define N_λ^{-} to be the class of distributions given by

$$N_\lambda^{-} = \{N^{-}(\lambda,\chi,\psi) : (\chi,\psi) \in \Theta_\lambda\}.$$

Let us introduce the parameters ω and η which are given by

$$\omega = \sqrt{\chi\psi}, \qquad \eta = \sqrt{\frac{\chi}{\psi}}.$$

It is convenient to let $\omega = 0$ denote the case where either $\chi = 0$, $\lambda > 0$ or $\psi = 0$, $\lambda < 0$. Thus $\omega > 0$ denotes the case where both χ and ψ are positive and where the density (1.1) takes the alternative form

$$\frac{\eta^{-\lambda}}{2K_\lambda(\omega)} x^{\lambda-1} e^{-\frac{1}{2}\omega(\eta x^{-1} + \eta^{-1}x)} \tag{2.2}$$

We use this notation rather freely, and in particular for $\omega = 0$ formula (2.2) is to be interpreted as (1.1) using the asymptotic formula for the norming constant mentioned above.

It follows from (2.2) that ω is a concentration parameter (for λ fixed) whereas η is a scale parameter. In the case $\omega = 0$ the parameter η has no meaning, but if $\chi = 0$, $\lambda > 0$ we have that ψ^{-1} is a scale parameter, and if $\psi = 0$, $\lambda < 0$ we have that χ is a scale parameter. In particular for any fixed (λ, ω) we have a scale parameter family. As λ is the index of the Bessel function in the norming constant of (1.1) we call λ the index parameter, or simply the index.

To indicate the analogy to the von Mises distribution, note that the last factor in (2.2) may be written in the form

$$e^{-\frac{1}{2}\omega \underline{\eta} \cdot \underline{x}},$$

where $\underline{x} = (x^{-1}, x)$ and $\underline{\eta} = (\eta, \eta^{-1})$ are vectors on the unit hyperbola in $\mathbb{R}^2$. This should be compared to the factor

$$e^{\kappa \underline{m} \cdot \underline{v}}$$

in the density of the von Mises distribution where $\underline{m}$ and $\underline{v}$ are vectors on the unit circle, $\underline{v}$ being the observation and κ and $\underline{m}$ being respectively the concentration parameter and the direction parameter. Note also the identity

$$(\chi, \psi) = \omega(\eta, \eta^{-1}) \tag{2.3}$$

which is the decomposition of the vector (χ, ψ) into hyperbolic coordinates, (η,η^{-1}) being the intersection point with the unit hyperbola (or the direction) and ω being the length of (χ,ψ) measured in units of (η,η^{-1}).

If the random variable X has distribution $N^{-}(\lambda,\chi,\psi)$ it is easily seen that

$$X^{-1} \sim N^{-}(-\lambda,\psi,\chi) \tag{2.4}$$

and if $c>0$ that

$$cX \sim N^{-}(\lambda,c\chi,c^{-1}\psi), \tag{2.5}$$

and thus ω is left unchanged after both scale and reciprocal transformations. Note that the class N_0^{-} is closed with respect to both kinds of transformations.

The density (1.1) is unimodal and the mode point is given by

$$m = \begin{cases} \dfrac{\lambda - 1 + \sqrt{(\lambda-1)^2 + \chi\psi}}{\psi} & \text{if } \psi > 0 \\[2ex] \dfrac{\chi}{2(1-\lambda)} & \text{if } \psi = 0. \end{cases} \tag{2.6}$$

It follows that the mode is positive except for the case $\chi = 0$, $0<\lambda\leq 1$. The distribution is strongly unimodal (i.e. has log-concave density) if $\lambda \geq 1$.

Figure 2.1 shows plots of the density (1.1) for a number of values for λ and ω. The scale parameter is chosen in such a way that the variance is unity. Some of the distributions with $\psi = 0$ have infinite variance (cf. section 2.1) and have consequently been excluded from the figure.

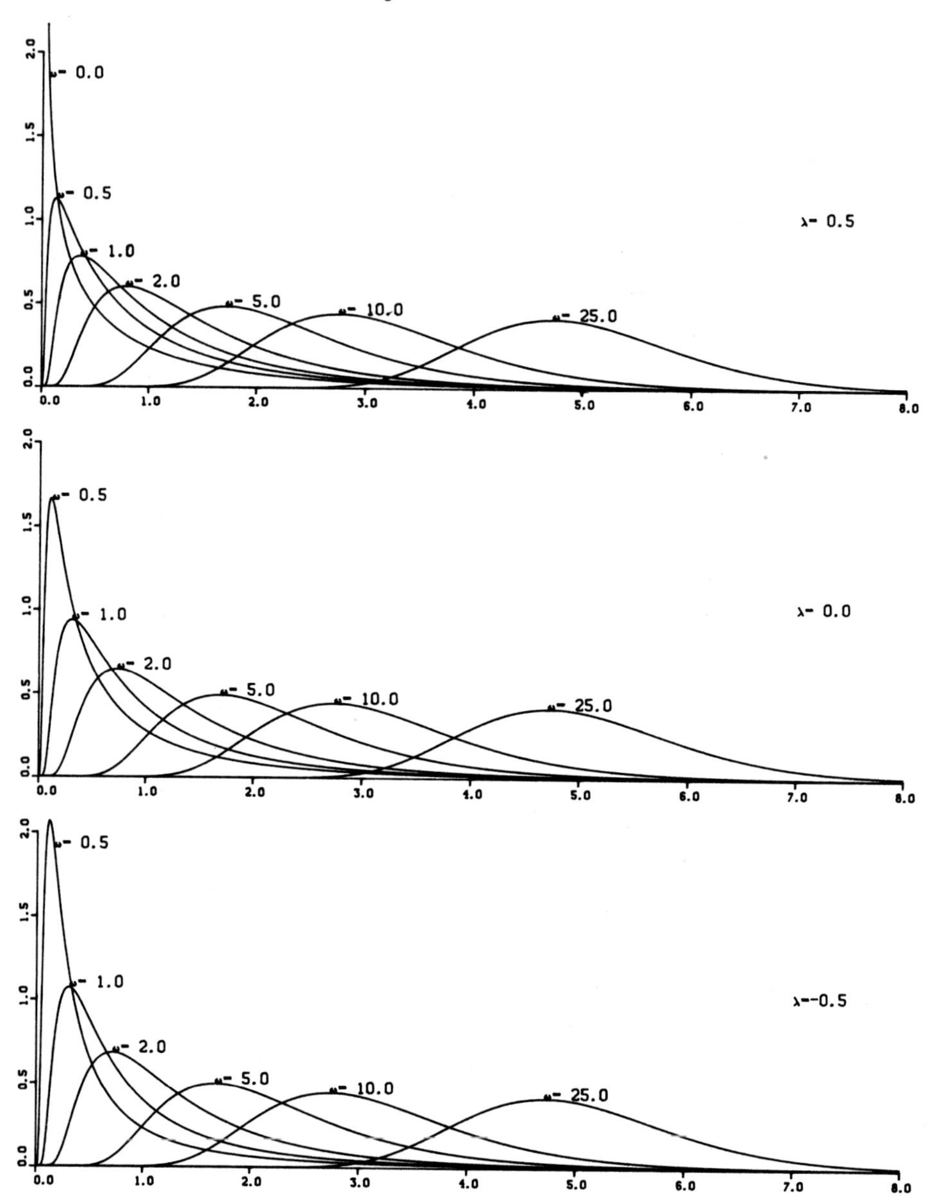

Figure 2.1. Plots of the probability density function (1.1). The values of λ and ω are indicated at each plot. The scale parameter has been chosen to make the variance unity. [Continued.]

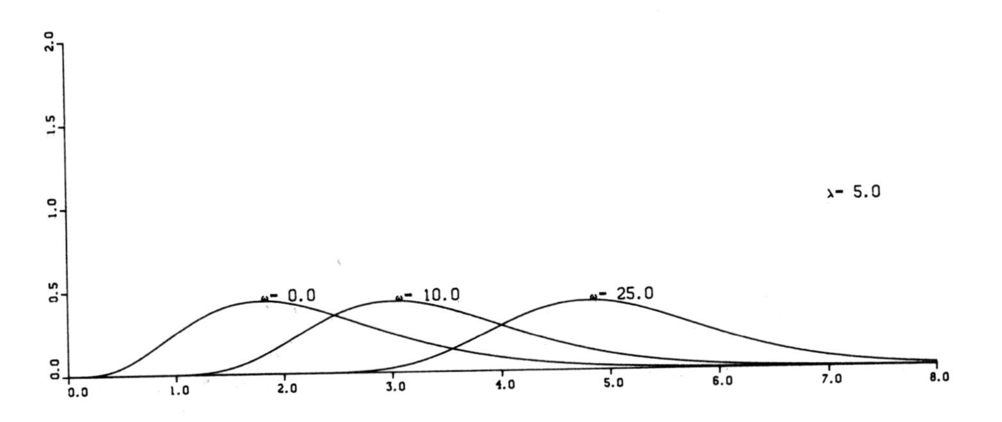

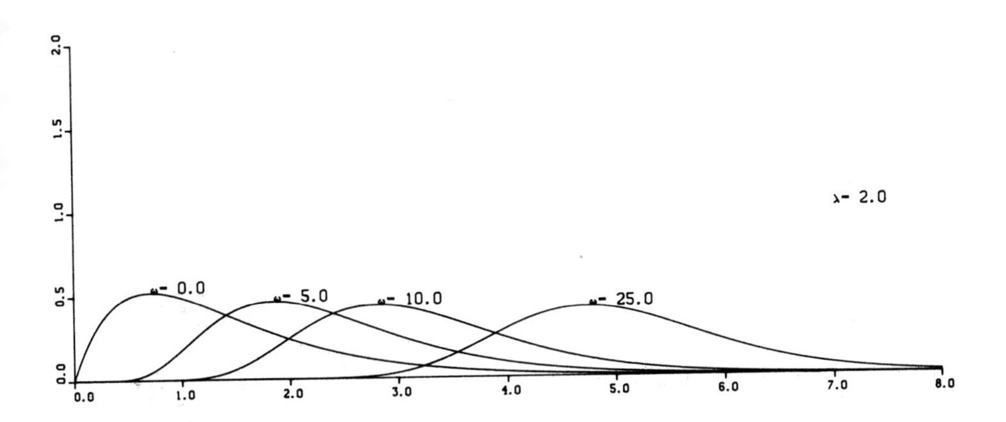

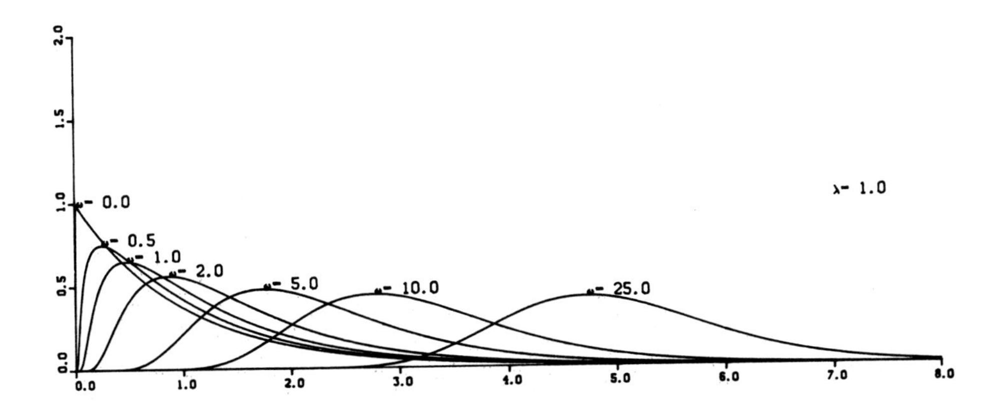

Figure 2.1. [Continued.]

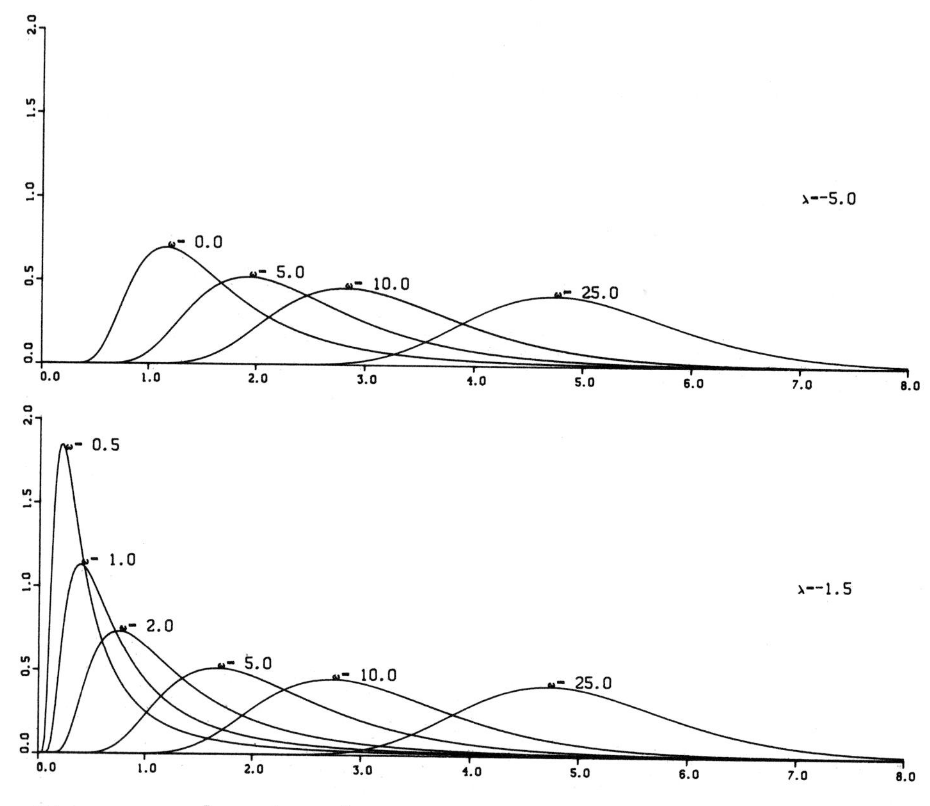

Figure 2.1. [Continued].

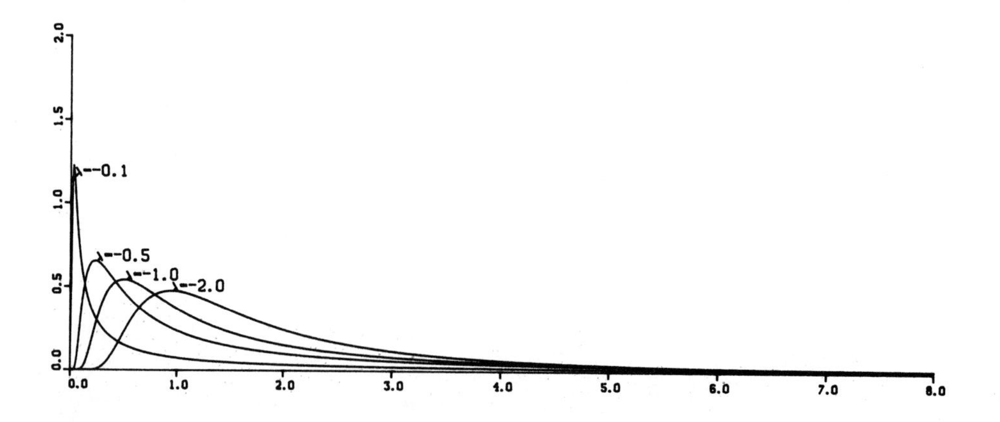

Figure 2.2. Plots of probability density function (1.1) for $\psi = 0$. The values of λ are indicated in the plot and the values of χ are given by $\chi = -2\lambda^{3/2}$.

Figure 2.2 shows plots of (1.1) with $\psi = 0$ for some values of λ where we have chosen $\chi = -2\lambda^{3/2}$. This value is reached at in the same manner as above by taking the "variance" of X to be $-\lambda^{-3}(\frac{\chi}{2})^2$ (cf. section (2.1)).

The class of generalized inverse Gaussian distributions is clearly a full exponential family of order 3 and by (2.1) not open and hence not regular. If $X_1,\ldots,X_n$ are independent random variables and $X_i \sim N^{-}(\lambda,\chi,\psi)$, the statistic

$$(X_{-},X_{\cdot},X_{\sim}) = \left(\sum_{i=1}^{n} X_i^{-1}, \sum_{i=1}^{n} X_i, \sum_{i=1}^{n} \ln X_i\right) \tag{2.7}$$

is minimal sufficient and complete. The notation given in (2.7) and the notation

$$\bar{X}_{-} = \frac{1}{n}X_{-}, \quad \bar{X}_{\cdot} = \frac{1}{n}X_{\cdot}, \quad \bar{X}_{\sim} = \frac{1}{n}X_{\sim}$$

is used throughout the work. Observed values of random variables are denoted by the corresponding lower case letters.

The cumulant transform of the vector $(-\frac{1}{2}X_{-},-\frac{1}{2}X_{\cdot}\,,\, X_{\sim})$ with respect to the measure $\mu\times\mu\times\nu$, where μ denotes the Lebesgue measure on $\mathbb{R}$ and where $\nu(dx) = x^{-1}\mu(dx)$, is given by

$$\begin{aligned}\kappa(\lambda,\chi,\psi) &= -\ln a(\lambda,\chi,\psi)^n \\ &= n[\tfrac{\lambda}{2}\ln\chi - \tfrac{\lambda}{2}\ln\psi + \ln 2K_\lambda(\sqrt{\chi\psi})],\end{aligned} \tag{2.8}$$

where $a(\lambda,\chi,\psi)$ is the norming constant in (1.1). It follows indicentally that the function

$$f(\lambda) = \ln K_\lambda(\omega)$$

is strictly convex for any given $\omega > 0$.

Tha Laplace transform of (1.1) is given by

$$\xi(t) = \frac{K_\lambda(\omega(1+\frac{2}{\psi}t)^{1/2})}{K_\lambda(\omega)(1+\frac{2}{\psi}t)^{\lambda/2}}. \tag{2.9}$$

In the cases $\chi = 0$, $\lambda > 0$ and $\psi = 0$, $\lambda < 0$ it follows from (A.1) and (A.7) that

$$\xi(t) = (1+\frac{2}{\psi}t)^{-\lambda} \qquad (\chi = 0,\ \lambda > 0) \tag{2.10}$$

and

$$\xi(t) = \frac{2K_\lambda(\sqrt{2\chi t})}{\Gamma(-\lambda)(\frac{\chi}{2}t)^{\lambda/2}} \qquad (\psi = 0,\ \lambda < 0). \tag{2.11}$$

From Barndorff-Nielsen (1978b) we have the following convolution formulas

$$N^{-}(-\tfrac{1}{2},\chi_1,\psi) + N^{-}(-\tfrac{1}{2},\chi_2,\psi) = N^{-}(-\tfrac{1}{2},(\sqrt{\chi_1}+\sqrt{\chi_2})^2,\psi) \tag{2.12}$$

$$N^{-}(-\lambda,\chi,\psi) + N^{-}(\lambda,0,\psi) = N^{-}(\lambda,\chi,\psi) \qquad (\lambda > 0) \tag{2.13}$$

$$N^{-}(-\tfrac{1}{2},\chi_1,\psi) + N^{-}(\tfrac{1}{2},\chi_2,\psi) = N^{-}(\tfrac{1}{2},(\sqrt{\chi_1}+\sqrt{\chi_2})^2,\psi), \tag{2.14}$$

and for the gamma distribution we have the well-known result

$$N^{-}(\lambda_1,0,\psi) + N^{-}(\lambda_2,0,\psi) = N^{-}(\lambda_1+\lambda_2,0,\psi). \tag{2.15}$$

The result (2.12) is the well-known convolution formula for the inverse Gaussian distribution. For $\psi = 0$ (2.12) concerns the stable distribution on $(0,\infty)$ with characteristic exponent 1/2.

2.1. Moments and cumulants

Let X be a random variable with distribution (1.1). The moments $\mu_k' = EX^k$ are easily seen to be given by

$$\mu_k' = \frac{K_{\lambda+k}(\omega)}{K_\lambda(\omega)}\eta^k, \qquad k \in \mathbb{R}. \tag{2.16}$$

In the case $\chi = 0$, $\lambda > 0$ we have by (A.7)

$$\mu_k' = \begin{cases} \frac{\Gamma(\lambda+k)}{\Gamma(\lambda)}\left(\frac{2}{\psi}\right)^k & \text{if } k > -\lambda \\ \infty & \text{if } k \leq -\lambda, \end{cases} \tag{2.17}$$

and in the case $\psi = 0$, $\lambda < 0$ we have by (A.7) and (A.1)

$$\mu_k' = \begin{cases} \frac{\Gamma(-\lambda-k)}{\Gamma(-\lambda)}\left(\frac{\chi}{2}\right)^k & \text{if } k < -\lambda \\ \infty & \text{if } k \geq -\lambda. \end{cases} \tag{2.18}$$

Many formulas in the following are simplified by using the functions R_λ and D_λ defined by

$$R_\lambda(\omega) = \frac{K_{\lambda+1}(\omega)}{K_\lambda(\omega)},$$

$$D_\lambda(\omega) = \frac{K_{\lambda+1}(\omega)K_{\lambda-1}(\omega)}{K_\lambda(\omega)^2},$$

respectively. A number of important results concerning these functions are listed in the Appendix.

From (2.16), (2.17) and (2.18) we have

$$EX = \begin{cases} R_\lambda(\omega)\eta & \text{if } \omega > 0 \\ \frac{2\lambda}{\psi} & \text{if } \chi = 0,\ \lambda > 0 \\ \frac{\chi}{2(-\lambda-1)} & \text{if } \psi = 0,\ \lambda < -1 \\ \infty & \text{if } \psi = 0,\ -1 \leq \lambda < 0 \end{cases} \tag{2.19}$$

and

$$EX^{-1} = \begin{cases} R_{-\lambda}(\omega)\eta^{-1} & \text{if } \omega > 0 \\ \frac{\psi}{2(\lambda-1)} & \text{if } \chi = 0,\ \lambda > 1 \\ \infty & \text{if } \chi = 0,\ 0 < \lambda \leq 1 \\ \frac{-2\lambda}{\chi} & \text{if } \psi = 0,\ \lambda < 0. \end{cases} \tag{2.20}$$

From (2.16) it follows that the variance of the distribution is given by

$$V(X) = \eta^2\left(\frac{K_{\lambda+2}(\omega)}{K_\lambda(\omega)} - \frac{K_{\lambda+1}(\omega)^2}{K_\lambda(\omega)^2}\right),$$

and using the definitions of R_λ and D_λ we have

$$V(X) = \eta^2 R_\lambda(\omega)^2 (D_{\lambda+1}(\omega) - 1). \tag{2.21}$$

The covariance matrix for (X^{-1}, X) is easily seen to be given by (using (A.1))

$$V(X^{-1}, X) = \begin{Bmatrix} R_{-\lambda}(\omega)^2\eta^{-2}(D_{\lambda-1}(\omega)-1) & 1 - D_\lambda(\omega) \\ 1 - D_\lambda(\omega) & \eta^2 R_\lambda(\omega)^2(D_{\lambda+1}(\omega)-1) \end{Bmatrix}. \tag{2.22}$$

Since (1.1) is a minimal representation it follows that (2.22) is non-singular for $(\chi, \psi) \in \operatorname{int} \Theta_\lambda$.

From (2.19) and (2.21) it follows that the coefficient of variation for the distribution is given by

$$C(X) = \sqrt{D_{\lambda+1}(\omega) - 1} \qquad (\omega > 0) \tag{2.23}$$

and using (A.20) we have

$$C(X) = \begin{cases} \lambda^{-1/2} & \text{if } \chi = 0,\ \lambda > 0 \\ (-\lambda-2)^{-1/2} & \text{if } \psi = 0,\ \lambda < -2 \\ \infty & \text{if } \psi = 0,\ -2 \leq \lambda < 0. \end{cases} \tag{2.24}$$

In chapter 4 (Theorem 4.1) we show that $D_\lambda(\omega)$ is decreasing as a function of ω and we have thus justified the use of the term concentration parameter about ω.

In particular it follows that for given $\lambda > 0$ $(\lambda < 0)$ the gamma (reciprocal gamma) distribution with index λ is characterized among the distributions in the class N_λ^{-1} as having the largest coefficient of variation. This parallels the characterization of the uniform distribution on the circle among the von Mises distributions. Note also that from Theorem A.1 (in the Appendix) we have that $C(X)$ is a decreasing (increasing) function of λ for $\lambda > -1$ $(\lambda < -1)$.

As η is a scale parameter we can express the j'th cumulant, $\varkappa_j$, in the form

$$\varkappa_j = W_{\lambda j}(\omega)\eta^j \qquad (\omega > 0) \tag{2.25}$$

and in the case $\omega = 0$ we have similarly

$$\begin{aligned} \varkappa_j &= W_{\lambda j}(0)\left(\tfrac{2}{\psi}\right)^j \qquad (\chi = 0,\ \lambda > 0) \\ \varkappa_j &= W_{\lambda j}(0)\left(\tfrac{\chi}{2}\right)^j \qquad (\psi = 0,\ \lambda < 0). \end{aligned} \tag{2.26}$$

This notation is convenient, but it should be noted that whereas the cumulants are right continuous at $\omega = 0$ this is not the case for the functions $W_{\lambda j}(\cdot)$.

We can find the functions $W_{\lambda j}$ by using the formulas expressing the cumulants in terms of the moments. Blæsild (1978) gives the following expressions for $W_{\lambda j}$, $j = 1,2,3$ and 4

$$\begin{aligned} W_{\lambda 1}(\omega) &= R_\lambda(\omega), \\ W_{\lambda 2}(\omega) &= -R_\lambda^2(\omega) + \frac{2(\lambda+1)}{\omega}R_\lambda(\omega) + 1, \\ W_{\lambda 3}(\omega) &= 2R_\lambda^3(\omega) - \frac{6(\lambda+1)}{\omega}R_\lambda^2(\omega) + \left(\frac{4(\lambda+1)(\lambda+2)}{\omega^2} - 2\right)R_\lambda(\omega) + \frac{2(\lambda+2)}{\omega}, \\ W_{\lambda 4}(\omega) &= -6R_\lambda^4(\omega) + \frac{24(\lambda+1)}{\omega}R_\lambda^3(\omega) + \left(\frac{-4(\lambda+1)(7\lambda+11)}{\omega^2} + 8\right)R_\lambda^2(\omega) \\ &\quad + \left(\frac{8(\lambda+1)(\lambda+2)(\lambda+3)}{\omega^3} - \frac{4(4\lambda+5)}{\omega}\right)R_\lambda(\omega) + \frac{4(\lambda+2)(\lambda+3)}{\omega^2} - 2. \end{aligned} \tag{2.27}$$

Note that this gives an alternative expression for the variance (cf. (2.21)).

In the case $\chi > 0$, $\lambda > 0$ we have

$$W_{\lambda j}(0) = \lambda(j-1)! \qquad j = 1,2,\ldots \tag{2.28}$$

and for $\psi = 0$, $\lambda < 0$

$$\begin{aligned} W_{\lambda 1}(0) &= \frac{1}{-\lambda-1} \qquad (\lambda < -1) \\ W_{\lambda 2}(0) &= \frac{1}{(-\lambda-1)^2(-\lambda-2)} \qquad (\lambda < -2) \end{aligned} \tag{2.29}$$

$$W_{\lambda 3}(0) = \frac{4}{(-\lambda-1)^3(-\lambda-2)(-\lambda-3)} \qquad (\lambda < -3)$$

$$W_{\lambda 4}(0) = \frac{-30\lambda-66}{(-\lambda-1)^4(-\lambda-2)^2(-\lambda-3)(-\lambda-4)} \qquad (\lambda < -4).$$

In the case $-j \le \lambda < 0$ we have $W_{\lambda j}(0) = \infty$.

The skewness and kurtosis of the distribution are given by

$$\gamma_{\lambda 1}(\omega) = \frac{W_{\lambda 3}(\omega)}{W_{\lambda 2}(\omega)^{3/2}},$$
$$\gamma_{\lambda 2}(\omega) = \frac{W_{\lambda 4}(\omega)}{W_{\lambda 2}(\omega)^{2}}, \qquad (2.30)$$

respectively, since these quantities are independent of scale parameters. In the case $\chi = 0$, $\lambda > 0$ we have from (2.28)

$$\gamma_{\lambda 1}(0) = 2\lambda^{-\frac{1}{2}}$$
$$\gamma_{\lambda 2}(0) = 6\lambda^{-1} \qquad (2.31)$$

and for $\psi = 0$, $\lambda < 0$ we have from (2.29)

$$\gamma_{\lambda 1}(0) = \begin{cases} \dfrac{4(-\lambda-2)^{1/2}}{-\lambda-3} & \text{if } \lambda < -3 \\ \infty & \text{if } -3 \le \lambda < 0 \end{cases}$$
$$\gamma_{\lambda 2}(0) = \begin{cases} \dfrac{-30\lambda-66}{(-\lambda-3)(-\lambda-4)} & \text{if } \lambda < -4 \\ \infty & \text{if } -4 \le \lambda < 0. \end{cases} \qquad (2.32)$$

From the right continuity of the cumulants at $\omega = 0$ it follows that

$$\lim_{\omega \downarrow 0} \gamma_{\lambda 1}(\omega) = \gamma_{\lambda 1}(0)$$
$$\lim_{\omega \downarrow 0} \gamma_{\lambda 2}(\omega) = \gamma_{\lambda 2}(0) \tag{2.33}$$

and from (2.27) and the zero order term of (A.21) it follows that

$$\lim_{\omega\to\infty} \gamma_{\lambda 1}(\omega) = \lim_{\omega\to\infty} \gamma_{\lambda 2}(\omega) = 0. \tag{2.34}$$

From (2.33) and (2.34) and numerical investigations Blæsild (1978) conjectured that the joint domain of variation for the skewness and kurtosis of the generalized inverse Gaussian distribution is simply the area bounded by the curves

$$\{(\gamma_{\lambda 1}(0), \gamma_{\lambda 2}(0)) : \lambda > 0\} \tag{2.35}$$

and

$$\{(\gamma_{\lambda 1}(0), \gamma_{\lambda 2}(0)) : \lambda < -4\}, \tag{2.36}$$

which are the curves of variation for the skewness and curtosis corresponding to the gamma and reciprocal gamma distributions, respectively. If Blæsild's conjecture is correct we have in particular that both the skewness and the kurtosis of the generalized inverse Gaussian distribution are positive.

In Figure 2.3 we have plotted the curves (2.35) and (2.36), but in terms of the traditional measures of skewness and kurtosis

$$(\beta_1, \beta_2) = (\gamma_1^2, \gamma_2 + 3).$$

Also in line with tradition we have reversed the secondary axis in Figure 2.3. Included in the plot are the curves corresponding to the inverse Gaussian and the log normal distributions.

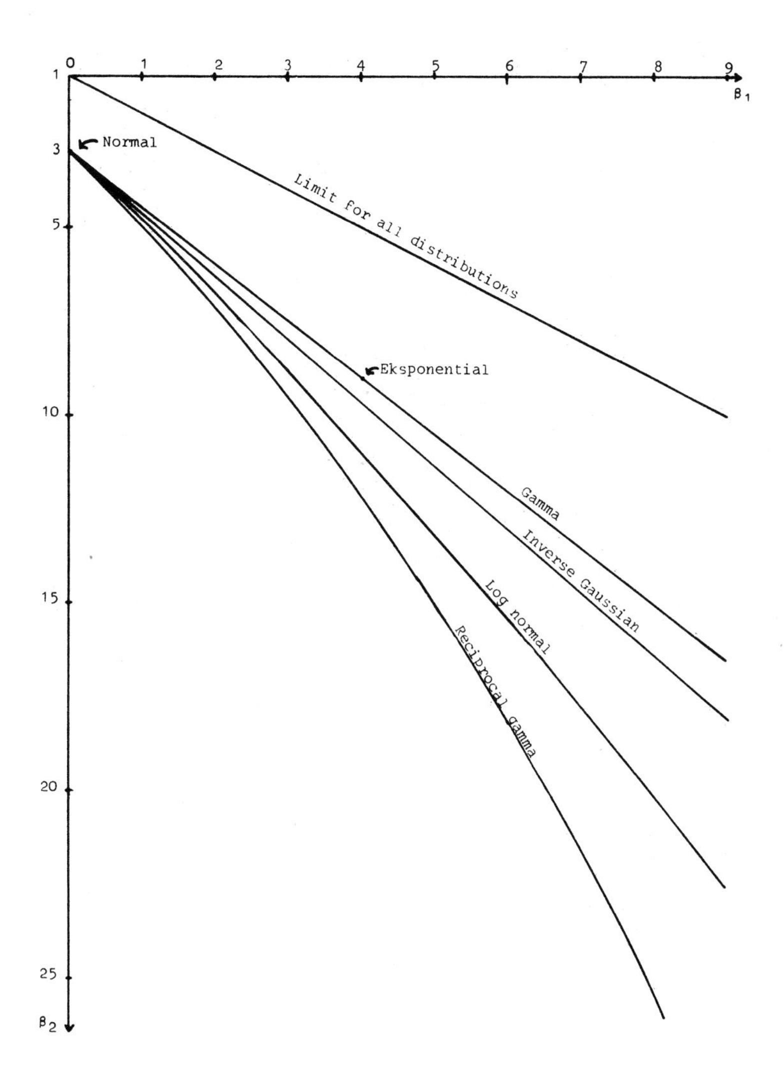

Figure 2.3. The asserted joint domain of variation for (β_1, β_2) for the generalized inverse Gaussian distribution (the area between the curves corresponding to the gamma and reciprocal gamma distributions).

3. Related distributions

In this chapter we consider a number of distributions with a certain relation to the generalized inverse Gaussian distribution.

In section 3.1 we show that the distribution is asymptotically normal for ω large and for $|\lambda|$ large.

In section 3.2 we consider the distribution of powers and of logarithms of a generalized inverse Gaussian variable and in section 3.3 we consider the distribution of products and quotients of independent generalized inverse Gaussian variates.

In section 3.4 we define a generalized inverse Gaussian Markov process, and we show that in some cases it has a simple stationary distribution. The subject may perhaps be said to fall outside the scope of the heading "related distributions", but we have put it in this chapter because the results are derived as a simple corrollary to the results in section 3.3.

In section 3.5 we consider the relation between the generalized inverse Gaussian and the generalized hyperbolic distributions.

3.1. Normal approximations

In this section we show that the distribution $N^{-}(\lambda,\chi,\psi)$ is asymptotically normal when ω tends to infinity for given λ, and when $|\lambda|$ tends to infinity for given ω. But before considering these results we derive some formulas for the moments of $\ln X$, where $X \sim N^{-}(\lambda,\chi,\psi)$.

For $\omega > 0$ the cumulant transform (2.8) (with $n = 1$) has the form

$$\kappa(\lambda,\chi,\psi) = \lambda \ln \eta + \ln 2 K_{\lambda}(\omega),$$

and by differentiating with respect to λ we have

$$E \ln X = \ln \eta + \frac{\partial}{\partial\lambda} \ln K_{\lambda}(\omega). \tag{3.1}$$

In the case $\lambda = 0$ we have

$$E \ln X = \ln \eta, \tag{3.2}$$

which incidentally follows by noting that in this case the distribution of $\ln X - \ln \eta$ is symmetric. Otherwise the derivative in (3.1) is non-zero.

In the case $\chi = 0$, $\lambda > 0$ the cumulant transform (2.8) is

$$\kappa(\lambda,0,\psi) = \ln \Gamma(\lambda) - \lambda \ln \frac{\psi}{2}$$

and it follows that

$$E \ln X = \Psi(\lambda) - \ln \frac{\psi}{2}, \tag{3.3}$$

where Ψ is the digamma function. Similarly one has for $\psi = 0$, $\lambda < 0$

$$E \ln X = -\psi(-\lambda) + \ln \frac{\chi}{2}. \tag{3.4}$$

We shall now derive asymptotic expansions for (3.1), (3.3) and (3.4).

From Abramowitz and Stegun (1965) we have the following asymptotic expansion of the digamma function for large values of λ

$$\psi(\lambda) = \ln \lambda - \frac{1}{2\lambda} - \frac{1}{12\lambda^2} + \frac{1}{120\lambda^4} - \frac{1}{252\lambda^6} + \ldots$$

Substituting into (3.3) and (3.4) one obtains asymptotic expansions of $E \ln X$ for $\omega = 0$, $|\lambda|$ large.

For an asymptotic expansion of (3.1) for large values of ω we turn to Laplace's method as described by Olver (1974) (chapter 8). One can obtain the following expansions

$$\int_1^\infty x^{\lambda-1} e^{-\frac{\omega}{2}(x^{-1}+x)} \ln x \, dx = e^{-\omega}\left[\frac{1}{\omega} + \sqrt{\frac{\pi}{2}} \frac{\lambda}{\omega^{3/2}} + \frac{3\lambda^2-1}{3\omega^2} + \ldots\right] \quad (\omega \to \infty) \tag{3.5}$$

and

$$\int_1^\infty x^{\lambda-1} e^{-\frac{\omega}{2}(x^{-1}+x)} (\ln x)^2 dx = e^{-\omega}\left[\frac{1}{2} \frac{\sqrt{2\pi}}{\omega^{3/2}} + \frac{2\lambda}{\omega^2} + \frac{3}{4}\sqrt{2\pi} \frac{\lambda^2 - \frac{5}{12}}{\omega^{5/2}} + \ldots\right] \quad (\omega \to \infty). \tag{3.6}$$

Using (A.9) we obtain from (3.5) and (2.2) (noting that $\ln \eta$ is a location parameter in the distribution of $\ln X$)

$$E \ln X = \ln \eta + \frac{\lambda}{\omega} + \ldots \qquad (\omega \to \infty). \tag{3.7}$$

For $\eta = 1$ we have from (3.6)

$$E(\ln X)^2 = \frac{1}{\omega} + \frac{\lambda^2 - \frac{1}{2}}{\omega^2} + \dots \qquad (\omega \to \infty), \tag{3.8}$$

and combining (3.7) (with $\eta = 1$) and (3.8) we have (using that $\ln \eta$ is a location parameter)

$$V \ln X = \frac{1}{\omega} - \frac{1}{2\omega^2} + \dots \qquad (\omega \to \infty). \tag{3.9}$$

We shall now consider asymptotic normality. We first consider the case where λ is fixed and ω tends to infinity. It is well-known that the inverse Gaussian distribution is asymptotically normal as ω tends to infinity (cf. Chhikara and Folks, 1978), and in fact we have for any λ and η

$$\sqrt{\omega}(X/\eta - 1) \stackrel{\sim}{\to} N(0,1) \qquad (\omega \to \infty) \tag{3.10}$$

$$\sqrt{\omega} \ln X/\eta \stackrel{\sim}{\to} N(0,1) \qquad (\omega \to \infty). \tag{3.11}$$

Note the analogy to the von Mises distribution which is also asymptotically normal as the concentration parameter tends to infinity. Note also that the asymptotic distribution is independent of λ.

To prove (3.10), note that from (2.2) the variable $\sqrt{\omega}(X/\eta - 1)$ has density function

$$\frac{\omega^{-1/2} e^{-\omega}}{2K_\lambda(\omega)} (1 + \omega^{-1/2}x)^{\lambda - 1} e^{-\frac{1}{2}\frac{x^2}{1 + \omega^{-1/2}x}} \qquad (x > -\omega^{1/2}) \tag{3.12}$$

and using (A.9) it follows that (3.12) converges pointwise to the standard normal density as ω tends to infinity. As the logarithm is totally differentiable (3.11) follows from (3.10).

Whitmore and Yalovsky (1978) have proved (3.11) for the inverse Gaussian distribution. They found that the distribution of the variate

$$\frac{1}{2\sqrt{\omega}} + \sqrt{\omega}\ln X/\eta \tag{3.13}$$

tends faster to normality than the distribution of $\sqrt{\omega}(X/\eta - 1)$ in (3.10). One notes that the correction term $1/2\sqrt{\omega}$ in (3.13) is obtained by approximating the mean of $\sqrt{\omega}\ln X/\eta$ using (3.7) (with $\lambda = -\frac{1}{2}$). Because X is a positive variate we believe that (3.11) tends faster to normality than does (3.10) for any λ, and below we examine the rate of convergence in (3.11) numerically. But before doing so we consider asymptotic normality as $|\lambda|$ tends to infinity.

It is well-known that the gamma distribution tends to normality as λ tends to infinity, that is for $\chi = 0$ we have

$$\frac{\psi}{2}\lambda^{-\frac{1}{2}}(X - \lambda\frac{2}{\psi}) \stackrel{\sim}{\to} N(0,1) \qquad (\lambda \to \infty) \tag{3.14}$$

and we shall prove that (3.14) holds for any $\chi \geq 0$. This simply follows by noting that for large λ the Laplace transform (2.9) is, using (A.10),

$$\xi(t) \simeq (1 + \frac{2}{\psi}t)^{-\lambda} \qquad (\lambda \to \infty)$$

which is the Laplace transform (2.10) of a gamma distribution. It follows that the asymptotic distribution is independent of χ, and hence (3.14) holds for any $\chi \geq 0$.

By noting that the function x^{-1} is totally differentiable it follows from (3.14) and (2.4) that

$$-\lambda^{3/2}\frac{2}{\chi}(X+\lambda^{-1}\frac{\chi}{2}) \stackrel{\sim}{\rightarrow} N(0,1) \qquad (\lambda\to-\infty) \qquad (3.15)$$

for any $\chi > 0$, $\psi \geq 0$.

A faster convergence to normality in (3.14) and (3.15) can (at least for the gamma distribution) be obtained by considering $\ln X$ instead of X and using asymptotic expansions for the mean and variance of $\ln X$ (cf. Cox and Lewis, 1966). Johnson and Kotz (1970) consider a number of approximations to the gamma distribution.

We have made a numerical investigation of the rate of convergence in (3.11). Figure 3.1 shows the log-density of the standardized variate $\tilde{u} = \sigma^{-1}(\ln X - \mu)$, where μ and σ^2 are the approximate mean and variance of $\ln X$, according to (3.7) and (3.9), for $\lambda = 0, 0.5, 1$ and $\omega = 2, 4, 10$, together with the log-density of the standard normal distribution. (Note that by (2.4) it is sufficient to consider the case $\lambda \geq 0$.)

We have found that the approximation to normality is fairly good in the central part of the distribution when ω is greater than $5(|\lambda|+1)$. However, since we also have convergence to normality as $|\lambda|$ tends to infinity it is obvious that for large $|\lambda|$ we might obtain a faster rate of convergence by using better approximations to the mean and variance. In particular we might use the exact mean and variance of $\ln X$, which may be obtained either by numerical integration or by numerical differentiation, cf. (3.1). A simple ad hoc solution is to take the values of μ and σ such that the density of $\sigma^{-1}(\ln X - \mu)$ has the correct mode point and mode value.

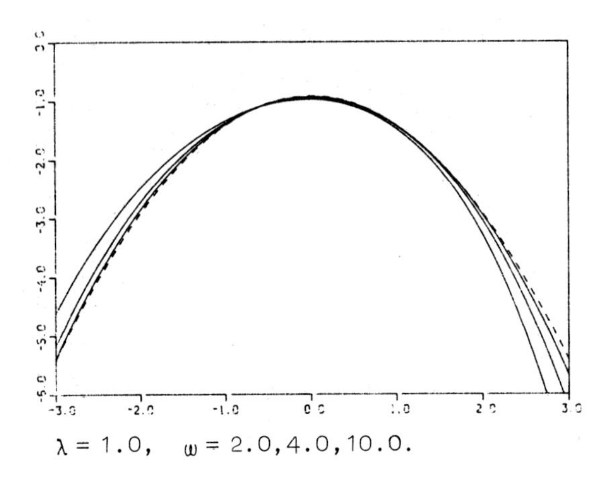

$\lambda = 1.0, \quad \omega = 2.0, 4.0, 10.0.$

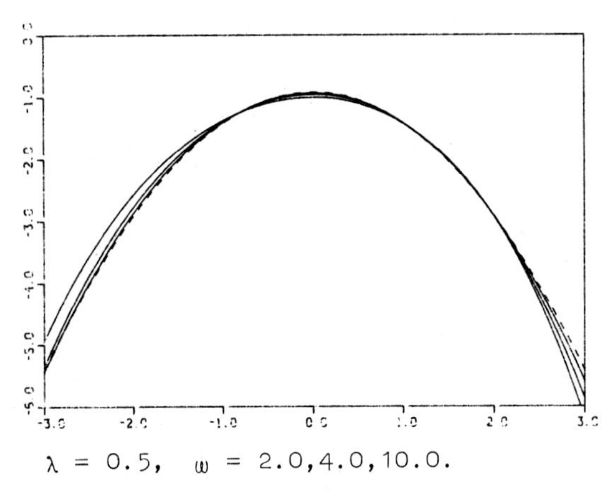

$\lambda = 0.5, \quad \omega = 2.0, 4.0, 10.0.$

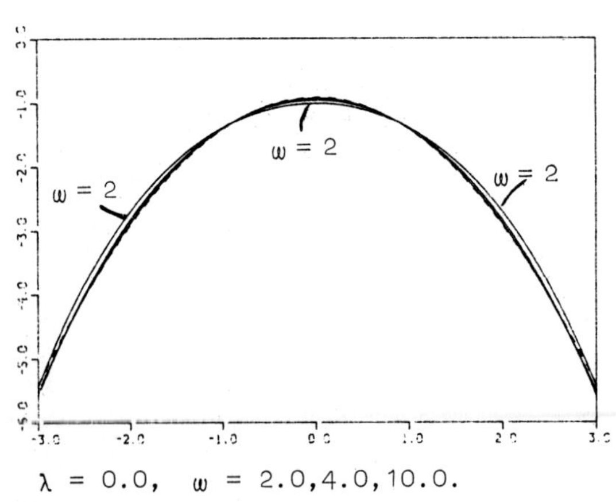

$\lambda = 0.0, \quad \omega = 2.0, 4.0, 10.0.$

Figure 3.1. Plots of log-density for a standardized log-generalized inverse Gaussian variate (see text) (solic curves) and log-density for a standard normal variate (dashed curve). On the two upper plots the density increases (decreases) with ω in the right (left) tail.

3.2. Powers and logarithms of generalized inverse Gaussian variates

For some purposes it would be natural to introduce a scale parameter in the log-scale, which corresponds to considering an arbitrary power of a generalized inverse Gaussian variate. The resulting distribution has probability density function

$$\frac{\alpha(\psi/\chi)^{\frac{\lambda}{2\alpha}}}{2K_{\frac{\lambda}{\alpha}}(\sqrt{\chi\psi})}x^{\lambda-1}e^{-\frac{1}{2}(\chi x^{-\alpha}+\psi x^{\alpha})}, \tag{3.16}$$

where λ, χ and ψ have the same domain of variation as before, and where $\alpha > 0$. If X has distribution (3.16) then $X^{\alpha} \sim N^{-}(\frac{\lambda}{\alpha}, \chi, \psi)$ so that properties for the generalized inverse Gaussian distribution are easily transformed to properties for (3.16).

Special cases of (3.16) are the generalized gamma distribution ($\chi = 0$, $\lambda > 0$) and the Weibull distribution ($\chi = 0$, $\lambda = \alpha$). In particular (3.16) includes distributions which are negatively skewed. From (3.11) we observe that the log normal distribution may be obtained as a limiting case of (3.16). Thus we find that (3.16) is a rather flexible class of distributions, which has of course 4 parameters.

One notes that when α is not fixed, (3.16) is not an exponential family.

Let us consider the distribution of $U = \ln X$, where X has distribution (3.16). In the case $\omega > 0$ the density for U may be written in the form

$$\frac{\alpha^{-1}}{2K_\lambda(\omega)} e^{\lambda\frac{u-\xi}{\alpha} - \omega \cosh\frac{u-\xi}{\alpha}} \qquad (u \in \mathbb{R}), \tag{3.17}$$

emphasizing that for fixed λ and ω we have a location and scale family. The logarithm of a generalized inverse Gaussian variate is for $\omega > 0$ distributed according to (3.17) with $\alpha = 1$ and $\xi = \ln \eta$.

The hyperbola distribution ($\lambda = 0$) was considered in the form (3.17) by both Rukhin (1974) and Barndorff-Nielsen (1978b), the former considering a general $\alpha > 0$ and the latter $\alpha = 1$.

In the case $\lambda \neq 0$ (3.17) has of course a valid limiting distribution for $\omega = 0$, namely that of the logarithm of a generalized gamma variate, but Rukhin (1974) noted that there is a limiting form of (3.17) for $\omega \downarrow 0$ also in the case $\lambda = 0$. Assuming $\xi = 0$ and letting $\omega \downarrow 0$ and $\alpha \downarrow 0$ in such a way that

$$-\alpha \ln \omega \to \delta > 0$$

we have, using (A.8),

$$\frac{\alpha^{-1}}{2K_0(\omega)} e^{-\omega \cosh\frac{u}{\alpha}} \simeq \frac{1}{2\delta} e^{-\frac{\omega}{2}(\omega^{u/\delta} + \omega^{-u/\delta})}$$

$$\underset{\omega \downarrow 0}{\to} \begin{cases} \frac{1}{2\delta} & \text{if } |u| < \delta \\ 0 & \text{if } |u| > \delta \end{cases}$$

which is the uniform distribution on $[-\delta, \delta]$.

3.3. Products and quotients of generalized inverse Gaussian variates

In many cases it is quite easy to find the distribution of quotients and products of generalized inverse Gaussian variates. We consider two examples with a certain relation to the hyperbola. The results of the first example are used in section 3.4 where we construct a generalized inverse Gaussian Markov process. We shall also consider the distribution of a variate which is essentially the exponent in the density (1.1).

Suppose that X_1, X_2 are independent and

$$X_i \sim N^{-}(\lambda_i, \chi_i, \psi_i), \quad i = 1,2,$$

and let $T = \sqrt{X_1 X_2}$, $S = \sqrt{X_2/X_1}$. Note that S and T are essentially the hyperbolic coordinates for the vector (X_1, X_2). The joint density for S and T is easily found to be (in an obvious notation)

$$\frac{\eta_1^{-\lambda_1}\eta_2^{-\lambda_2}}{2K_{\lambda_1}(\omega_1)K_{\lambda_2}(\omega_2)} t^{\lambda_1+\lambda_2-1} s^{\lambda_2-\lambda_1-1} e^{-\frac{1}{2}(s^{-1}(\psi_1 t+\chi_2 t^{-1})+s(\chi_1 t^{-1}+\psi_2 t))}$$

$$(t,s>0), \quad (3.18)$$

and hence

$$S|T = t \sim N^{-}(\lambda_2-\lambda_1, \psi_1 t+\chi_2 t^{-1}, \chi_1 t^{-1}+\psi_2 t). \quad (3.19)$$

One notes that any generalized inverse Gaussian distribution may appear in (3.19). Even if $\chi_1 = \chi_2 = 0$ (i.e. if X_1 and X_2 have gamma distributions) any generalized inverse Gaussian distribution with $\omega > 0$ may appear in (3.19), and if $\psi_1 = \chi_2 = 0$ (i.e. if X_2 and X_1^{-1} have gamma distributions) we have that (3.19) is a gamma distribution. It should also be noted that (3.19) depends on (λ_1, λ_2) only through $\lambda_2 - \lambda_1$, in particular (3.19) is a hyperbola distribution for $\lambda_1 = \lambda_2$.

Fisher (1956, p. 165-171) illustrates the ideas of ancillarity, recovery of information and fiducial inference by an example where X_1 and X_2 are independent random variables having gamma distributions whose means belong to a hyperbola (i.e. $\lambda_1 = \lambda_2 > 0$, $\chi_1 = \chi_2 = 0$, $\psi_1 = \psi_2^{-1} = \psi$). Fisher derived the conditional distribution of the maximum likelihood estimator $\hat{\psi} = \sqrt{X_2/X_1}$ given the ancillary statistic $T = \sqrt{X_1 X_2}$, a special case of (3.19) which is in fact a hyperbola distribution, since $\lambda_1 = \lambda_2$. See also Efron and Hinkley (1978).

By integrating out t in (3.18) we obtain the following density for s

$$\frac{\eta_1^{-\lambda_1} \eta_2^{-\lambda_2}}{K_{\lambda_1}(\omega_1) K_{\lambda_2}(\omega_2)} s^{\lambda_2 - \lambda_1 - 1} K_{\lambda_1+\lambda_2}((s^{-1}\chi_2 + s\chi_1)^{\frac{1}{2}}(s^{-1}\psi_1 + s\psi_2)^{\frac{1}{2}})$$

$$\times \left(\frac{s^{-1}\chi_2 + s\chi_1}{s\psi_2 + s^{-1}\psi_1}\right)^{\frac{\lambda_1+\lambda_2}{2}} \qquad (s > 0). \qquad (3.20)$$

Since s^2 is a quotient between two independent generalized inverse Gaussian variates we may view (3.20) as a generalization of the F-distribution.

Note that the distribution of a product of two generalized inverse Gaussian variates may be obtained from (3.20) by using (2.4), and in a similar way we may obtain the conditional distribution of $T|S = s$ from (3.19).

For the second example, let $X_1,\ldots,X_n$ be independent and $X_i \sim N^{-}(\lambda,\chi,\psi)$, $i = 1,\ldots,n$. Let us find the distribution of the vector

$$Y = (Y_1,\ldots,Y_n),$$

where

$$Y_i = X_i/X_{i+1}, \quad i = 1,\ldots,n-1,$$

$$Y_n = X_n.$$

Note here that $(Y_1,\ldots,Y_{n-1})$ is maximal invariant with respect to the group of scale transformations.

To find the distribution of Y, note that Y has domain of variation $\mathbb{R}_+^n$ and that the inverse transformation is given by

$$X_i = \prod_{k=i}^{n} Y_k, \quad i = 1,\ldots,n .$$

Hence the Jacobian of the inverse transformation is $\prod_{i=1}^{n} y_i^{i-1}$ and it follows that Y has density

$$\frac{\eta^{-n\lambda}}{2^n K_\lambda(\omega)^n} \prod_{i=1}^{n} y_i^{i\lambda-1} e^{-\frac{1}{2}(\chi \sum_{k=1}^{n} \prod_{i=k}^{n} y_i^{-1} + \psi \sum_{k=1}^{n} \prod_{i=k}^{n} y_i)} \tag{3.21}$$

$$(y_1,\ldots,y_n > 0).$$

Inspection of (3.21) shows that the conditional distribution of Y_k given the remaining $n-1$ variables is generalized inverse Gaussian with index $k\lambda$. In particular we have

$$Y_n | Y_1 = y_1,\dots,Y_{n-1} = y_{n-1} \sim N^{-}(n\lambda, \chi \sum_{k=1}^{n} \prod_{i=k}^{n-1} y_i^{-1}, \psi \sum_{k=1}^{n} \prod_{i=k}^{n-1} y_i). \qquad (3.22)$$

If we integrate (3.21) with respect to y_n we obtain the following density for $Y_1,\dots,Y_{n-1}$.

$$\frac{K_{n\lambda}(\omega(\sum_{k=1}^{n} \prod_{i=k}^{n-1} y_i^{-1})^{\frac{1}{2}}(\sum_{k=1}^{n} \prod_{i=k}^{n-1} y_i)^{\frac{1}{2}})}{2^{n-1} K_\lambda(\omega)^n} \prod_{i=1}^{n-1} y_i^{i\lambda-1} \left(\frac{\sum_{k=1}^{n} \prod_{i=k}^{n-1} y_i^{-1}}{\sum_{k=1}^{n} \prod_{i=k}^{n-1} y_i} \right)^{\frac{n\lambda}{2}}. \qquad (3.23)$$

One notes that $\sqrt{X.X_{-}}$ (cf. (2.7)) is a function of $Y_1,\dots,Y_{n-1}$, in fact

$$\sqrt{X.X_{-}} = (\sum_{k=1}^{n} \prod_{i=k}^{n-1} Y_i^{-1})^{\frac{1}{2}} (\sum_{k=1}^{n} \prod_{i=k}^{n-1} Y_i)^{\frac{1}{2}},$$

a term which appears in (3.23). However it seems not in general feasible to obtain the distribution of $\sqrt{X.X_{-}}$ by integration of (3.23). We return to the distribution of $\sqrt{X.X_{-}}$ in section 5.1 where we find it in some special cases.

Finally, let us consider the distribution of the variate

$$Z = \frac{\chi(X-\eta)^2}{\eta^2 X}$$

$$= \chi X^{-1} + \psi X - 2\sqrt{\chi\psi}, \qquad (3.24)$$

where $X \sim N^{-}(\lambda,\chi,\psi)$. The Laplace transform of Z is given by

$$E(e^{-Zt}) = e^{2t\omega} \frac{K_\lambda(\omega(1+2t))}{K_\lambda(\omega)}. \qquad (3.25)$$

For the inverse Gaussian distribution we have by (A.4)

$$E(e^{-Zt}) = (1+2t)^{-\frac{1}{2}} \tag{3.26}$$

which is the Laplace transform of a chi-squared distribution with one degree of freedom (Shuster, 1968).

Traditionally, the density of the inverse Gaussian distribution is taken to have $-\frac{1}{2}z$ in the exponent, so that the negative of twice the term in the exponent is a chi-squared variable, just as for the normal distribution.

For $\lambda = \frac{1}{2}$ the distribution of Z is the same as for $\lambda = -\frac{1}{2}$ and if $\omega = 0$ we have by (A.7)

$$E(e^{-Zt}) = (1+2t)^{-|\lambda|},$$

which is the Laplace transform of a gamma distribution (this also follows directly from (3.24)). From (A.9) one has that (3.26) is the limit of (3.25) as ω tends to infinity.

In all other cases Z does not have a chi-squared distribution, a fact which helps to explain the unique character of the inverse Gaussian distribution.

3.4. A generalized inverse Gaussian Markov process

We shall now define a Markov process where the conditional distribution governing the process is generalized inverse Gaussian, and we consider briefly the question of finding a stationary distribution for the process.

Let $U = \{U_i: i = 0,1,...\}$ be a Markov process defined by the conditional distribution (3.19), i.e.

$$U_{i+1}|U_i = u \sim N^{-}(\lambda_2-\lambda_1,\psi_1 u+\chi_2 u^{-1},\chi_1 u^{-1}+\psi_2 u), \quad i=1,2,... \quad (3.27)$$

and by specifying the distribution of U_0. Thus for each i, U_{i+1} plays the role of S and U_i plays the role of T in (3.19). (For simplicity we follow the notation of section 3.3.) Since (3.27) depends on (λ_1,λ_2) only through $\lambda_2-\lambda_1$ we take $\lambda_1 = 0$.

Assume from now on that $\chi_1 = \psi_1 = \omega$. We shall show that in this case we can find a stationary distribution for the process by a simple argument.

Recall from section 3.3 that the variables S and T are defined by

$$S = \sqrt{X_2/X_1}, \qquad T = \sqrt{X_1 X_2},$$

where X_1 and X_2 are independent, $X_2 \sim N^{-}(\lambda_2,\chi_2,\psi_2)$ and where now $X_1 \sim N^{-}(0,\omega,\omega)$. From (2.4) it follows that X_1 and X_1^{-1} have the same distribution, and hence that S and T have the same distribution. Having defined the conditional distribution (3.27) to be the same as that of $S|T$ it follows trivially that the common marginal distribution for S and T

is a stationary distribution for the process U. Thus, by (3.20) the density for this stationary distribution is

$$f(u) = \frac{\eta_2^{-\lambda_2}}{K_0(\omega)K_{\lambda_2}(\omega_2)} u^{\lambda_2 - 1} K_{\lambda_2}((\omega u + \chi_2 u^{-1})^{\frac{1}{2}}(\omega u^{-1} + \psi_2 u)^{\frac{1}{2}}) \left(\frac{\omega u + \chi_2 u^{-1}}{\omega u^{-1} + \psi_2 u}\right)^{\frac{\lambda_2}{2}}$$

$$(u > 0). \qquad (3.28)$$

Let us examine some special cases of (3.28). For $\chi_2 = \psi_2 = \omega$ we have

$$f(u) = \frac{u^{\lambda_2 - 1}}{K_0(\omega)K_{\lambda_2}(\omega)} K_{\lambda_2}(\omega(u + u^{-1})) \qquad (3.29)$$

whereas the density corresponding to (3.27) turns into

$$f(x|u) = \frac{1}{2K_{\lambda_2}(\omega(u + u^{-1}))} x^{\lambda_2 - 1} e^{-\frac{1}{2}\omega(u + u^{-1})(x + x^{-1})}. \qquad (3.30)$$

If we instead assume $\lambda_2 = \frac{1}{2}$ we obtain, cf.(A.4),

$$f(u) = \frac{\sqrt{\psi_2}\, e^{\omega_2} e^{-(\omega u + \chi_2 u^{-1})^{\frac{1}{2}}(\omega u^{-1} + \psi_2 u)^{\frac{1}{2}}}}{K_0(\omega)\, u^{\frac{1}{2}} (\omega u^{-1} + \psi_2 u)^{\frac{1}{2}}} \qquad (3.31)$$

$$f(x|u) = \frac{\sqrt{\omega u^{-1} + \psi_2 u}}{\sqrt{2\pi}} e^{(\omega u + \chi_2 u^{-1})^{\frac{1}{2}}(\omega u^{-1} + \psi_2 u)^{\frac{1}{2}}}$$

$$\times\, x^{-\frac{1}{2}} e^{-\frac{1}{2}((\omega u + \chi_2 u^{-1})x^{-1} + (\omega u^{-1} + \psi_2 u)x)}. \qquad (3.32)$$

For $\lambda_2 = -\frac{1}{2}$ we obtain

$$f(u) = \frac{\sqrt{\chi_2}\, e^{\omega_2} e^{-(\omega u + \chi_2 u^{-1})^{\frac{1}{2}}(\omega u^{-1} + \psi_2 u)^{\frac{1}{2}}}}{K_0(\omega)\ u^{\frac{3}{2}}\ (\omega u + \chi_2 u^{-1})^{\frac{1}{2}}} \tag{3.33}$$

and

$$f(x|u) = \frac{\sqrt{\omega u + \chi_2 u^{-1}}}{\sqrt{2\pi}}\, e^{(\omega u + \chi_2 u^{-1})^{\frac{1}{2}}(\omega u^{-1} + \psi_2 u)^{\frac{1}{2}}}$$

$$\times\ x^{-\frac{3}{2}} e^{-\frac{1}{2}((\omega u + \chi_2 u^{-1})x^{-1} + (\omega u^{-1} + \psi_2 u)x)} . \tag{3.34}$$

In the case where also $\chi_2 = \psi_2 = \omega$, (3.31), (3.32), (3.33) and (3.34) turn into respectively

$$f(u) = \frac{e^{\omega} e^{-\omega(u+u^{-1})}}{K_0(\omega) u^{\frac{1}{2}} (u+u^{-1})^{\frac{1}{2}}} ,$$

$$f(x|u) = \frac{\sqrt{\omega(u+u^{-1})}}{\sqrt{2\pi}}\, e^{\omega(u+u^{-1})} x^{-\frac{1}{2}} e^{-\frac{1}{2}\omega (u+u^{-1})(x+x^{-1})} ,$$

$$f(u) = \frac{e^{\omega} e^{-\omega(u+u^{-1})}}{K_0(\omega) u^{\frac{3}{2}} (u+u^{-1})^{\frac{1}{2}}}$$

and

$$f(x|u) = \frac{\sqrt{\omega(u+u^{-1})}}{\sqrt{2\pi}}\, e^{\omega(u+u^{-1})} x^{-\frac{3}{2}} e^{-\frac{1}{2}\omega(u+u^{-1})(x+x^{-1})} .$$

In the formulas above for the stationary distribution there have obviously not appeared any standard distributions.

3.5. The generalized hyperbolic distribution

Suppose that the mean ξ and the variance σ^2 of a normal distribution are related by the equation $\xi = \mu + \beta\sigma^2$. If σ^2 follows the distribution $N^{-}(\lambda,\delta^2,\alpha^2-\beta^2)$ the resulting mixture distribution is the (one dimensional) generalized hyperbolic distribution whose probability density function is given by

$$\frac{\sqrt{\alpha^2-\beta^2}^{\lambda}}{\sqrt{2\pi}\,\alpha^{\lambda-\frac{1}{2}}\delta^{\lambda}K_{\lambda}(\delta\sqrt{\alpha^2-\beta^2})}\sqrt{\delta^2+(x-\mu)^2}^{\lambda-\frac{1}{2}}K_{\lambda-\frac{1}{2}}(\alpha\sqrt{\delta^2+(x-\mu)^2})e^{\beta(x-\mu)}$$

$$(x \in \mathbb{R}). \qquad (3.35)$$

For uses of this distribution, see Barndorff-Nielsen (1977, 1978b). We show that (1.1) may be obtained as a limiting case of (3.35).

If $\varphi = \alpha+\beta$, $\gamma = \alpha-\beta$ and $\mu = 0$ (3.35) takes the form

$$\frac{\sqrt{\varphi\gamma}^{\lambda}(\delta^2+x^2)^{\lambda-\frac{1}{2}}}{\sqrt{2\pi}\,(\frac{\varphi+\gamma}{2})^{\lambda-\frac{1}{2}}\delta^{\lambda}K_{\lambda}(\delta\sqrt{\varphi\gamma})}K_{\lambda-\frac{1}{2}}(\tfrac{\varphi+\gamma}{2}\sqrt{\delta^2+x^2})e^{\frac{\varphi-\gamma}{2}x}. \qquad (3.36)$$

If we let $\varphi \to \infty$ and $\delta \to 0$ in such a way that $\varphi\delta^2 \to c$ it follows by using (A.9) that (3.36) converges to the density corresponding to the distribution $N^{-}(\lambda,\frac{c}{2},2\gamma)$.

In the case $\lambda = 0$, $\mu = 0$, $\delta = 1$ the density (3.35) takes the form

$$\frac{1}{2K_0(\omega)\sqrt{1+x^2}}e^{-\alpha\sqrt{1+x^2}+\beta x}, \qquad (3.37)$$

where $\omega = \sqrt{\alpha^2+\beta^2}$. If we apply the variate transformation

$$x = \sinh u$$

the density (3.37) turns into

$$\frac{1}{2K_0(\omega)} e^{-\alpha \cosh u + \beta \sinh u}. \tag{3.38}$$

Taking $\eta = \sqrt{\frac{\alpha+\beta}{\alpha-\beta}}$ it easily follows that (3.38) and (3.17) (with $\lambda = 0$, $\alpha = 1$ and $\xi = \ln \eta$) are in fact identical, and hence (3.38) is one form of the density of the hyperbola distribution. This curious fact was observed by Barndorff-Nielsen (1978b).

4. Maximum likelihood estimation

This chapter deals with maximum likelihood estimation based on n independent observations $X_1,\dots,X_n$ from the distribution $N^{-}(\lambda,\chi,\psi)$.

In section 4.1 we discuss estimation of (χ,ψ) when λ is fixed and in section 4.2 we consider the asymptotic distribution of the estimate.

Section 4.3 concerns the estimation of λ and centers on the properties of the partially maximized log-likelihood for λ.

The sections 4.4 and 4.5 deal briefly with estimation in the cases where λ and η or λ and ψ, respectively, are fixed.

Throughout we use terminology and results of Barndorff-Nielsen (1978a) concerning exponential families.

4.1. Estimation for fixed λ

In this section we consider maximum likelihood estimation in the family $N_\lambda^{\dashv}$, i e. estimation of (χ,ψ) when λ is fixed. The estimation result is given in Theorem 4.1 below, but first we introduce the likelihood equations and give some introductory comments.

The likelihood equations have the form

$$R_\lambda(\omega)\eta = \bar{x}. \tag{4.1a}$$

$$R_{-\lambda}(\omega)\eta^{-1} = \bar{x}_{\dashv} \tag{4.1b}$$

(cf. (2.19) and (2.20)). If we consider the ratio and the product of (4.1a) and (4.1b) we get the alternative equations (using (A.14))

$$D_\lambda(\omega) = \bar{x}.\bar{x}_{\dashv} \tag{4.2}$$

$$\eta = \sqrt{\frac{\bar{x}.}{\bar{x}_{\dashv}}}\sqrt{\frac{R_{-\lambda}(\omega)}{R_\lambda(\omega)}}. \tag{4.3}$$

Thus if the likelihood equations have a solution it may be found by solving (4.2) for ω and then inserting in (4.3) to get η, and it follows that the properties of D_λ are central in the discussion of the estimation (cf. (A.20) and (A.22)).

Note here that

$$\bar{x}_{\dashv}^{-1} \leq \exp(\bar{x}_{\sim}) \leq \bar{x}. , \tag{4.4}$$

and hence

$$\bar{x}.\bar{x}_{\dashv} \geq 1, \tag{4.5}$$

since the variables in (4.4) are respectively the harmonic, geometric and arithmetic mean, and note that equality in (4.4) (and (4.5)) occurs if and only if $x_1 = \ldots = x_n$. We are now ready to give the estimation result.

Theorem 4.1. (Estimation for fixed λ.) Assume that $\bar{x}.\bar{x}_{-1} > 1$. Then the maximum likelihood estimate $(\hat{\chi}_\lambda, \hat{\psi}_\lambda)$ exists and is unique, and is given as follows:

In the case $|\lambda| \leq 1$ the family N_λ^{-1} is steep (in particular N_0^{-1} is regular), and the estimate is the unique solution to the likelihood equations (4.1).

In the case $|\lambda| > 1$ the family is not steep. For $\bar{x}.\bar{x}_{-1} < |\lambda|/(|\lambda| - 1)$ the estimate is the unique solution to the likelihood equations (4.1). In the opposite case $\bar{x}.\bar{x}_{-1} \geq |\lambda|/(|\lambda|-1)$ the estimate is given by

$$(\hat{\chi}_\lambda, \hat{\psi}_\lambda) = \begin{cases} (0, 2\lambda/\bar{x}.) & (\lambda > 1) \qquad (4.6a) \\ (-2\lambda/\bar{x}_{-1}, 0) & (\lambda < -1), \qquad (4.6b) \end{cases}$$

which in the case $\lambda > 1$ $(\lambda < -1)$ corresponds to maximum likelihood estimation for the gamma (reciprocal gamma) distribution.

If $\bar{x}.\bar{x}_{-1} = 1$ the likelihood does not attain its supremum.

The function $D_\lambda(\cdot)$ is strictly decreasing and maps the interval $(0,\infty)$ onto the interval

$$I = \begin{cases} (1,\infty) & \text{if } |\lambda| \leq 1 \\ (1, |\lambda|/(|\lambda|-1) & \text{if } |\lambda| > 1. \end{cases}$$

**

The proof of the theorem is deferred to the end of this section, and we shall now give some comments on the estimate, and we shall consider some approximations.

Note that if we define $u = \bar{x}.\bar{x}_{-1}/(\bar{x}.\bar{x}_{-1} - 1)$ we have from (4.5) that $u > 1$, and hence by the properties of D_λ we see that the likelihood equations (or equivalently (4.2) and (4.3)) have a solution if and only if $|\lambda| < u$, and in the opposite case the estimate is given by (4.6).

In general the equation (4.2) can only be solved numerically, but as noted in the Appendix D_λ is a rational function when $\lambda + \frac{1}{2}$ is an integer. However it seems only feasible to solve (4.2) explicitly in the cases $|\lambda| = \frac{1}{2}$ and $|\lambda| = \frac{3}{2}$.

For $|\lambda| = \frac{1}{2}$ we have from (A.18) and (4.2)

$$\hat{\omega}_{\pm\frac{1}{2}} = \frac{1}{\bar{x}.\,\bar{x}_{-1} - 1}, \tag{4.7}$$

which for the inverse Gaussian distribution yields the well-known result (Tweedie, 1957)

$$\hat{\eta}_{-\frac{1}{2}} = \bar{x}. \qquad \hat{\chi}^{-1}_{-\frac{1}{2}} = \bar{x}_{-1} - \bar{x}.^{-1}. \tag{4.8}$$

In the general case the equation (4.2) is easily solved iteratively on a computer, though this of course requires access to a routine that calculates K_λ. We have found that the function $d(x) = \ln(D_\lambda(e^x)-1$ is well suited for Newton-Raphson iteration (the required derivative of D_λ is given by (A.24)). Figure A.2 in the Appendix shows plots of D_λ for some values of λ, and Figure 4.1 shows plots of the function d, or rather log-log plots of $D_\lambda(\omega) - 1$, for some values of λ.

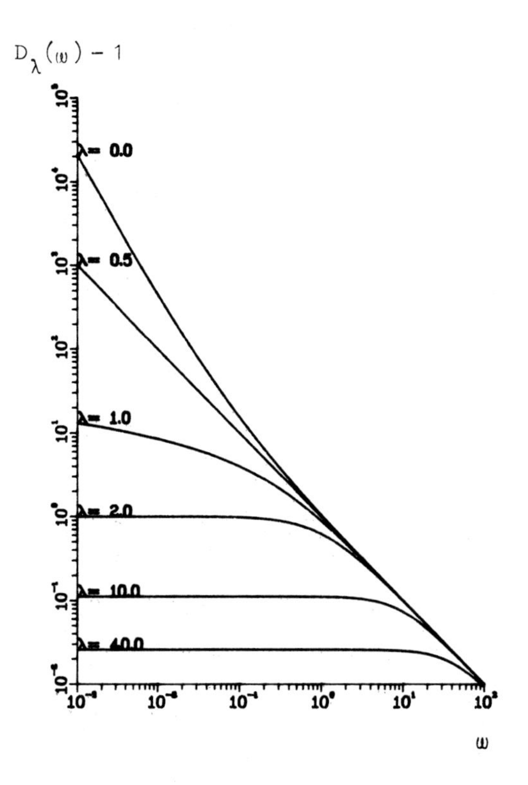

Figure 4.1. Log-log plots of the function $D_\lambda(\omega) - 1$ for some values of λ.

Let us now give some interpretation of the estimate.

From (4.2) it follows that $\hat{\omega}_\lambda$ is a function of $\bar{x}.\bar{x}_{-1}$ only, and is of course invariant under scale transformations. One would also expect $\hat{\omega}_\lambda$ to be a measure of the concentration of the observations, which is in fact the case since D_λ is decreasing and $\bar{x}.\bar{x}_{-1}$ is a measure of the dispersion of the observations. Note in this connection that the estimated coefficient of variation, $(D_{\lambda+1}(\hat{\omega}_\lambda) - 1)^{\frac{1}{2}}$, (cf. (2.23)) is a decreasing function of $\hat{\omega}_\lambda$, and hence an increasing function of $\bar{x}.\bar{x}_{-1}$.

In the case $|\lambda| > 1$ we have that $\hat{\omega}_\lambda = 0$ if $\bar{x}.\bar{x}_\dashv > |\lambda|/(|\lambda| - 1)$. This means that if the dispersion of the observations is too large, the estimator points at the gamma or the reciprocal gamma (depending on the sign of λ) as being the most likely distribution. In section 2.1 we saw that for a given $\lambda > 0$ $(\lambda < 0)$ the gamma (reciprocal gamma) distribution has maximum coefficient of variation among the distributions in the family $N_\lambda^{\dashv}$.

In the case $\lambda = 0$ the estimate (4.3) for η has the particularly simple form

$$\hat{\eta}_0 = \sqrt{\frac{\bar{x}.}{\bar{x}_\dashv}} .$$

As noted by Barndorff-Nielsen (1978b) the point $(\hat{\eta}_0^{-1}, \hat{\eta}_0)$ is the intersection point between the unit hyperbola and the resultant vector $(x_\dashv, x.)$, showing a close analogy to the von Mises distribution in this case.

For $\lambda \neq 0$ the estimate for η contains an additional factor which depends on $\hat{\omega}_\lambda$ through a quotient between Bessel functions. Using (4.1), (4.2) and (A.17) we may obtain an expression which does not involve the Bessel functions, namely

$$\hat{\eta}_\lambda = \frac{-\lambda/\hat{\omega}_\lambda + \sqrt{(\lambda/\hat{\omega}_\lambda)^2 + \bar{x}.\bar{x}_\dashv}}{\bar{x}_\dashv} = \frac{\bar{x}.}{\lambda/\hat{\omega}_\lambda + \sqrt{(\lambda/\hat{\omega}_\lambda)^2 + \bar{x}.\bar{x}_\dashv}} \qquad (4.9)$$

This is in fact the maximum likelihood estimate for η when $\omega = \hat{\omega}_\lambda$ is given.

Finally we give some approximations for $D_\lambda(\omega)$ when ω is large or small.

From (A.22) we have

$$D_\lambda(\omega) \simeq 1 + \frac{1}{\omega} \qquad (\omega \to \infty) \tag{4.10}$$

and hence

$$\hat{\omega}_\lambda \simeq \frac{1}{\bar{x}.\bar{x}_{\dashv} - 1} \qquad (\bar{x}.\bar{x}_{\dashv} \downarrow 1) \tag{4.11}$$

(compare with (4.7)).

For $\lambda = 0$ we get from (A.7) and (A.8)

$$D_0(\omega) \simeq (\omega \ln \omega)^{-2} \qquad (\omega \downarrow 0) \tag{4.12}$$

and for $\lambda = 1$

$$D_1(\omega) \simeq 2 \ln \omega^{-1} \qquad (\omega \downarrow 0).$$

For $0 < |\lambda| < 1$ we obtain from (A.7)

$$D_\lambda(\omega) \simeq |\lambda| \frac{\Gamma(1-|\lambda|)}{\Gamma(|\lambda|)} 4^{1-|\lambda|} \omega^{2(|\lambda|-1)} \qquad (\omega \downarrow 0). \tag{4.13}$$

Using (A.5), (A.6) and (A.7) we have for $1 < |\lambda| < 2$

$$D_\lambda(\omega) \simeq \frac{|\lambda|}{|\lambda|-1}\left(1 - \frac{\Gamma(2-|\lambda|)}{\Gamma(|\lambda|)} 4^{1-|\lambda|} \omega^{2(|\lambda|-1)}\right) \qquad (\omega \downarrow 0) \tag{4.14}$$

and for $|\lambda| > 2$

$$D_\lambda(\omega) \simeq \frac{|\lambda|}{|\lambda|-1}\left(1 - \frac{\omega^2}{2|\lambda|(|\lambda|-1)(|\lambda|-2)}\right) \qquad (\omega \downarrow 0). \tag{4.15}$$

In the case $|\lambda| = 2$ we get from Abramowitz and Stegun (1965, 9.6.53)

$$D_2(\omega) \simeq 2\left(1 + \omega^2\left(\tfrac{3}{8} + \tfrac{1}{2} \ln \tfrac{1}{2}\omega\right)\right) \qquad (\omega \downarrow 0). \tag{4.16}$$

From these results we may obtain approximations for $\hat{\omega}_\lambda$, and we shall make use of the approximations later.

Proof of Theorem 4.1

Note that the family N_λ^{-} is a full exponential family of order 2 with (χ,ψ) as minimal canonical parameter and $(t_1,t_2) = (-\frac{1}{2}x_{\rightarrow}, -\frac{1}{2}x.)$ as minimal canonical statistic. Since the family is full it is in particular convex, and in Barndorff-Nielsen (1978a, chapter 9) the estimation problem for a general convex exponential family is solved. Hence we only need to spell out the result as it looks in the present case. First some notation.

The log-likelihood function takes the form

$$l(\lambda,\chi,\psi) = -\kappa_\lambda(\chi,\psi) + (\lambda-1)x_\sim + \chi t_1 + \psi t_2, \tag{4.17}$$

where $\kappa_\lambda(\chi,\psi) = \kappa(\lambda,\chi,\psi)$ is the cumulant transform (2.8). Let τ_λ denote the mean value mapping

$$\tau_\lambda(\chi,\psi) = (\tau_{\lambda 1}(\chi,\psi),\tau_{\lambda 2}(\chi,\psi))$$

$$= E_{\lambda,\chi,\psi}(T_1,T_2).$$

From (2.19) and (2.20) we have

$$\tau_\lambda(\chi,\psi) = (-\frac{n}{2}R_{-\lambda}(\omega)\eta^{-1},-\frac{n}{2}R_\lambda(\omega)\eta) \quad (\omega>0). \tag{4.18}$$

Letting $\mathfrak{J}_\lambda = \tau_\lambda(\operatorname{int}\Theta_\lambda)$ one has $\mathfrak{J}_\lambda \subseteq \operatorname{int} C$, where C is the convex support for (T_1,T_2). From (4.5) we have

$$\operatorname{int} C = \{(t_1,t_2): t_1<0, t_1t_2>\frac{n^2}{4}\}.$$

The maximum likelihood estimate $(\hat{\chi}_\lambda, \hat{\psi}_\lambda)$ exists (and is unique) if and only if $(t_1, t_2) \in \operatorname{int} C$, i.e. if and only if $\bar{x}.\bar{x}_{-1} > 1$. In the opposite case the likelihood does not attain its supremum.

Figure 4.2 illustrated the set C and some other sets that we consider later.

From now on we assume $\lambda \geq 0$, since the opposite case follows from (2.4) by considering the reciprocal observations.

The estimation result depends on whether κ_λ is steep or not, and hence we examine this. Steepness means that the norm $|D\kappa_\lambda|$ (where $D\kappa_\lambda$ is the gradient of κ_λ) tends to infinity at any boundary point of Θ_λ. Now since

$$\frac{\partial}{\partial \chi}\kappa_\lambda(\chi,\psi) = \tau_{\lambda 1}(\chi,\psi), \quad \frac{\partial}{\partial \psi}\kappa_\lambda(\chi,\psi) = \tau_{\lambda 2}(\chi,\psi)$$

we have

$$|D\kappa_\lambda(\chi,\psi)|^2 = \tau_{\lambda 1}(\chi,\psi)^2 + \tau_{\lambda 2}(\chi,\psi)^2, \tag{4.19}$$

and hence steepness is implied if at least one of the terms in (4.19) tends to infinity at a given boundary point.

First, recall from (2.1) that

$$\Theta_0 = \{(\chi,\psi): \chi > 0, \ \psi > 0\}$$

and, for $\lambda > 0$,

$$\Theta_\lambda = \{(\chi,\psi): \chi \geq 0, \psi > 0\}.$$

Since Θ_0 is open we have that N_0^{-} is regular, in particular steep.

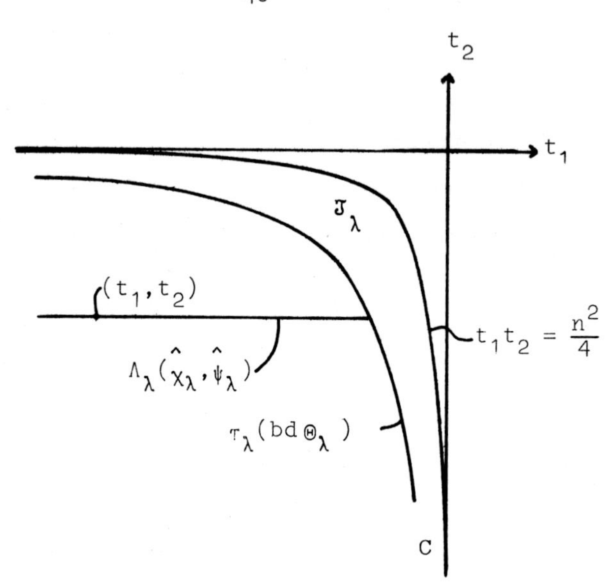

Figure 4.2. The sets C, $\mathcal{J}_\lambda$ and Λ_λ in the case $\bar{x}.\bar{x}_{-1} > \lambda/(\lambda-1)$.

Now consider the full exponential family of order 1 obtained by taking a fixed $\chi \geq 0$ and letting ψ vary. Clearly this family is regular, whence the corresponding cumulant transform $\kappa_\lambda(\chi,\cdot)$ is steep, and hence $\tau_{\lambda 2}(\chi,\psi)^2 \to \infty$ $(\psi \downarrow 0)$.

From (2.20) we get

$$\lim_{\chi \downarrow 0} \tau_{\lambda 1}(\chi,\psi) = \begin{cases} \dfrac{-n\psi}{4(\lambda-1)} & \text{if } \lambda > 1 \\ -\infty & \text{if } 0 < \lambda \leq 1, \end{cases} \tag{4.20}$$

and thus we have that κ_λ is steep for $0 \leq \lambda \leq 1$, and not steep for $\lambda > 1$.

In the case $0 \leq \lambda \leq 1$ the estimation is simple, since for a steep exponential family the maximum likelihood estimate is the unique solution to the likelihood equation

$$\tau_\lambda(\chi,\psi) = (t_1,t_2)$$

which is equivalent to (4.1).

Before we consider the non-steep case $\lambda > 1$ we shall prove that D_λ is monotone. Consider for a given $\omega > 0$ the set

$$\mathfrak{J}_\lambda(\omega) = \{\tau_\lambda(\chi,\psi) : \sqrt{\chi\psi} = \omega\}.$$

From (4.18) and (A.14) we have

$$\mathfrak{J}_\lambda(\omega) = \{(\tau_1,\tau_2) : \tau_1 < 0, \tau_1\tau_2 = \frac{n^2}{4} D_\lambda(\omega)\},$$

which is a hyperbola. Since τ_λ is one-to-one on $\operatorname{int}\Theta_\lambda$ these hyperbolas must be distinct, and since τ_λ is continuous on $\operatorname{int}\Theta_\lambda$ we conclude that $D_\lambda(\cdot)$ is monotone for any given λ. From (A.20) and the zero order term of (A.22) we find that D_λ is in fact decreasing and maps $(0,\infty)$ onto $(1,\infty)$ in the case $0 \leq \lambda \leq 1$ and onto $(1,\frac{\lambda}{\lambda-1})$ in the case $\lambda > 1$.

Now let us consider estimation in the non-steep case $\lambda > 1$. From the results just obtained we find

$$\mathfrak{J}_\lambda = \{(\tau_1,\tau_2) : \tau_1 < 0, \frac{n^2}{4} < \tau_1\tau_2 < \frac{n^2}{4}\frac{\lambda}{\lambda-1}\},$$

which is a proper subset of $\operatorname{int} C$, and

$$\tau_\lambda(\operatorname{bd}\Theta_\lambda) = \{(\tau_1,\tau_2) : \tau_1 < 0, \tau_1\tau_2 = \frac{n^2}{4}\frac{\lambda}{\lambda-1}\}$$

(see figure 4.2). It turns out that we have an example of a full exponential family where $\mathfrak{J}$ is not convex.

From Barndorff-Nielsen (1978a, p.160) we have that the mapping inverse to the maximum likelihood estimator is

$$\Lambda_\lambda(\chi,\psi) = \tau_\lambda(\chi,\psi) + M_\lambda(\chi,\psi), \tag{4.21}$$

where M_λ is the normal cone mapping for the set Θ_λ,

$$M_\lambda(\chi,\psi) = \{z \in \mathbb{R}^2 : ((\chi,\psi) - w)\cdot z \leq 0 \quad \forall\, w \in \Theta_\lambda\}.$$

We have

$$M_\lambda(\chi,\psi) = \begin{cases} 0 & \text{if } (\chi,\psi) \in \operatorname{int}\Theta_\lambda \\ \{k(-1,0) : k \in [0,\infty)\} & \text{if } \chi = 0 \end{cases}$$

and thus

$$\Lambda_\lambda(\chi,\psi) = \tau_\lambda(\chi,\psi) \qquad (\chi > 0) \tag{4.22}$$

$$\Lambda_\lambda(0,\psi) = \{\tau_\lambda(0,\psi) + k(-1,0) : k \in [0,\infty)\}. \tag{4.23}$$

Figure 4.2 illustrates the sets Λ_λ, $\mathfrak{J}_\lambda$ and C.

For $1 < \bar{x}.\bar{x}_{-1} < \lambda/(\lambda-1)$ we see that $(t_1,t_2) \in \mathfrak{J}_\lambda$, and hence from (4.22) that $(\hat{\chi}_\lambda, \hat{\psi}_\lambda)$ is the solution to the likelihood equation (4.1).

In the opposite case $\bar{x}.\bar{x}_{-1} \geq \lambda/(\lambda-1)$ we have $(t_1,t_2) \in \operatorname{int} C \setminus \mathfrak{J}_\lambda$ and it now follows from (4.23) (see figure 4.2) that $\hat{\chi}_\lambda = 0$ and that $\hat{\psi}_\lambda$ is the solution to the equation $\tau_{\lambda 2}(\chi,\psi) = t_2$ which gives (4.6a) by using (2.19). This clearly coincides with the maximum likelihood estimate for ψ (λ fixed) in the gamma distribution.

We have thus completed the proof. **

4.2. On the asymptotic distribution of the maximum likelihood estimate for fixed λ.

In this section we consider the asymptotic distribution of the maximum likelihood estimate $(\hat{\chi}_\lambda,\hat{\psi}_\lambda)$, as given in Theorem 4.1.

Let $i_\lambda(\chi,\psi) = V(-\frac{1}{2}X^{-1},-\frac{1}{2}X)$ denote the Fisher information for a single observation, which by (2.22) is

$$i_\lambda(\chi,\psi) = \frac{1}{4}\begin{Bmatrix} R_{-\lambda}(\omega)^2\eta^{-2}(D_{\lambda-1}(\omega)-1) & 1-D_\lambda(\omega) \\ 1-D_\lambda(\omega) & R_\lambda(\omega)^2\eta^2(D_{\omega+1}(\omega)-1) \end{Bmatrix}. \quad (4.24)$$

By the remark right after (2.22) we have that i_λ is non-singular for $\omega > 0$.

Note that for $\hat{\omega}_\lambda > 0$ the expression for the observed information may be simplified using (4.2) and the likelihood equation (4.1).

In the case $\omega > 0$ we have $(\chi,\psi) \in \operatorname{int}\Theta_\lambda$, and hence from standard asymptotic theory for exponential families

$$\sqrt{n}[(\hat{\chi}_\lambda,\hat{\psi}_\lambda)-(\chi,\psi)] \overset{\sim}{\to} N(0,i_\lambda(\chi,\psi)^{-1}) \quad (n \to \infty). \quad (4.25)$$

In Theorem 4.3 we consider the asymptotic distribution of the estimate in the case $\omega = 0$, but first we consider consistency of the estimate.

Theorem 4.2. For any $(\chi,\psi) \in \Theta_\lambda$ we have

$$(\hat{\chi}_\lambda,\hat{\psi}_\lambda) \to (\chi,\psi) \quad (n\to\infty) \quad \text{a.s.} \quad ** \quad (4.26)$$

Proof. The case $\omega > 0$ follows from standard asymptotic theory for exponential families.

Hence let us assume $\omega = 0$, $\lambda > 0$ (the case $\omega = 0$, $\lambda < 0$

then follows using (2.4)).

From the strong law of large numbers we have

$$(\bar{X}_{\dashv},\bar{X}.) \to E(X^{-1},X) \quad \text{a.s.,} \tag{4.27}$$

and hence, almost surely,

$$\bar{X}_{\dashv}\bar{X}. \to \begin{cases} \lambda/(\lambda-1) & \text{if } \lambda>1 \\ \infty & \text{if } 0<\lambda\leq 1, \end{cases}$$

where we have used (2.19) and (2.20). From Theorem 4.1 and the continuity of D_λ it follows by using (4.2) that $\hat{\omega}_\lambda \to 0$ almost surely.

Using now (4.1a) and (A.7) we have for small values of $\hat{\omega}_\lambda$

$$\hat{\psi}_\lambda = \bar{X}.^{-1}\, R_\lambda(\hat{\omega}_\lambda)\hat{\omega}_\lambda \simeq \bar{X}.^{-1}\, 2\lambda.$$

From (4.27) and (2.19) we have $\bar{X}. \to \frac{2\lambda}{\psi}$ almost surely, and hence $\hat{\psi}_\lambda \to \psi$ almost surely, establishing (4.26). **

We shall now consider the asymptotic distribution of the estimate, assuming that $\omega = 0$. In this case the standard theory no longer applies, and in particular it turns out that the asymptotic distribution of the estimate is no longer normal.

The results in Theorem 4.3 below are derived using the central limit theorem for the variables $(\bar{X}_{\dashv},\bar{X}.)$. In the case $\omega = 0$, $0<|\lambda|\leq 2$ the standard version of the central limit theorem, however, does not apply, because then either X or X^{-1} has infinite variance (for $0<|\lambda|\leq 1$ even infinite mean). Hence we have here restricted our attention to the case $|\lambda|>2$, but we are working on the case $|\lambda|\leq 2$. For $|\lambda|=\frac{1}{2}$ the exact distribution of the estimate is known (cf. section 5.1).

Theorem 4.3. Assume that $\omega = 0$. For $|\lambda| > 2$ we have

$$\sqrt{n}\,\hat{\omega}_{\lambda}^{2} \overset{\sim}{\rightarrow} \max(0,V) \quad (n\rightarrow\infty), \tag{4.28}$$

where $V \sim N(0,8|\lambda|(|\lambda|-1)^2(|\lambda|-2))$.

In the case $\lambda > 2$ we have the following results: The exact conditional distribution of $\hat{\psi}_{\lambda}$ given $\hat{\omega}_{\lambda}$ is

$$\hat{\psi}_{\lambda}|\hat{\omega}_{\lambda} = w \sim N^{-1}(-n\lambda, n\psi(\lambda+\sqrt{\lambda^2+w^2}D_{\lambda}(w)), 0). \tag{4.29}$$

For large n, $\hat{\psi}_{\lambda}$ has the following approximate density:

$$f_{\hat{\psi}_{\lambda}}(x) = \frac{1}{2}\frac{(n\lambda\psi)^{n\lambda}}{\Gamma(n\lambda)}x^{-n\lambda-1}e^{-\frac{n\lambda\psi}{x}}$$

$$+ \frac{\Gamma(n\lambda+\frac{2n\lambda}{\lambda-2})(\frac{n\psi}{2})^{n\lambda}(\frac{n}{\lambda-2})^{\frac{2n\lambda}{\lambda-2}}}{\Gamma(n\lambda)\Gamma(\frac{2n\lambda}{\lambda-2})(\frac{n\psi}{2x}+\frac{n}{\lambda-2})^{n\lambda+\frac{2n\lambda}{\lambda-2}}}x^{-n\lambda-1}\,\Gamma(4\lambda(\frac{n\psi}{2x}+\frac{n}{\lambda-2})|n\lambda+\frac{2n\lambda}{\lambda-2})$$

$$(x > 0), \tag{4.30}$$

where $\Gamma(\cdot)$ is the ordinary gamma function, whereas $\Gamma(\cdot|\lambda)$ denotes the upper tail area of the distribution $N^{-1}(\lambda,0,1)$, i.e. an incomplete gamma function.

The moments of the approximate distribution (4.30) are

$$E\hat{\psi}_{\lambda}^{k} = \frac{1}{2}\frac{\Gamma(n\lambda-k)}{\Gamma(n\lambda)}(n\lambda\psi)^{k}$$

$$+ \frac{\Gamma(n\lambda-k)\Gamma(k+\frac{2n\lambda}{\lambda-2})}{\Gamma(n\lambda)\Gamma(\frac{2n\lambda}{\lambda-2})}\left(\frac{(\lambda-2)\psi}{2}\right)^{k}\Gamma(\frac{4n\lambda}{\lambda-2}\,|\,k+\frac{2n\lambda}{\lambda-2}) \quad (k < n\lambda), \tag{4.31}$$

and in particular

$$E\hat{\psi}_\lambda = \frac{n\lambda\psi}{n\lambda-1}\left[\frac{1}{2}+\Gamma\left(\frac{4n\lambda}{\lambda-2}\,\middle|\,1+\frac{2n\lambda}{\lambda-2}\right)\right], \tag{4.32}$$

$$E\hat{\psi}_\lambda^2 = \frac{n\lambda\psi^2}{2(n\lambda-1)(n\lambda-2)}\left[n\lambda+(\lambda-2+2n\lambda)\Gamma\left(\frac{4n\lambda}{\lambda-2}\,\middle|\,2+\frac{2n\lambda}{\lambda-2}\right)\right]. \tag{4.33}$$

In the case $\lambda < -2$, $\omega = 0$ the above results apply if λ, ψ and $\hat{\psi}_\lambda$ are replaced by respectively $-\lambda$, χ and $\hat{\chi}_\lambda$.

Remark. For the gamma distribution we have the estimate $\tilde{\psi}_\lambda = 2\lambda/\bar{x}.$, which has distribution $N^{-1}(-n\lambda, n\psi 2\lambda, 0)$ and moments $(n\psi\lambda)^k\Gamma(n\lambda-k)/\Gamma(n\lambda)$. The results (4.30) - (4.33) are distorted versions of this, as a result, largely, of integrating out w in (4.29), using (4.28).

Proof. From (2.4) it follows that it is sufficient to consider the case $\lambda > 2$, $\omega = 0$.

From the central limit theorem we have, using (2.19) and (2.20),

$$\sqrt{n}\left[(\bar{X}_{-1}, \bar{X}.) - \left(\frac{\psi}{2(\lambda-1)}, \frac{2\lambda}{\psi}\right)\right] \tilde{\to} N(0, 4i_\lambda(0,\psi)) \quad (n\to\infty), \tag{4.34}$$

where, from (2.26), (2.28) and (2.29) and (A.20),

$$4i_\lambda(0,\psi) = \begin{Bmatrix} \dfrac{\psi^2}{4(\lambda-1)^2(\lambda-2)} & \dfrac{1}{1-\lambda} \\ \dfrac{1}{1-\lambda} & \dfrac{4\lambda}{\psi^2} \end{Bmatrix}.$$

From (4.34) we get

$$\sqrt{n}[\bar{X}.\bar{X}_{\dashv}\,\frac{\lambda-1}{\lambda} - 1] \overset{\sim}{\rightarrow} N(0,\frac{2}{\lambda(\lambda-2)}) \qquad (n\rightarrow\infty), \tag{4.35}$$

because the function $(\bar{X}_{\dashv},\bar{X}.) \rightarrow \bar{X}.\bar{X}_{\dashv}$ is totally differentiable (below we use similar arguments without special notice).

For $\hat{\omega}_\lambda$ small we get from (4.15) and Theorem 4.1

$$\frac{\hat{\omega}_\lambda^2}{2\lambda} \simeq \max(0,V_0),$$

where

$$V_0 = -(\lambda-1)(\lambda-2)[\bar{X}.\bar{X}_{\dashv}\,\frac{\lambda-1}{\lambda} - 1],$$

and using (4.35) we have

$$\sqrt{n}\;V_0 \overset{\sim}{\rightarrow} N(0,\tfrac{2}{\lambda}(\lambda-1)^2(\lambda-2)) \qquad (n\rightarrow\infty),$$

which shows (4.28).

From Basu's theorem we have that $\bar{X}.\bar{X}_{\dashv}$ and $\bar{X}.$ are independent, since for fixed λ, $\bar{X}.\bar{X}_{\dashv}$ is ancillary and $\bar{X}.$ sufficient and complete, and from (2.15) we have

$$X. \sim N^{\dashv}(n\lambda,0,\psi). \tag{4.36}$$

The following expression for $\hat{\psi}_\lambda$ may be obtained from (4.9)

$$\hat{\psi}_\lambda = \frac{\hat{\omega}_\lambda}{\hat{\eta}_\lambda} = \bar{X}.^{-1}\,(\lambda + \sqrt{\lambda^2 + \hat{\omega}_\lambda^2\;\bar{X}.\bar{X}_{\dashv}}), \tag{4.37}$$

which holds also for $\hat{\omega}_\lambda = 0$. From (4.37) one sees that $\hat{\psi}_\lambda$ is stochastically larger than the corresponding estimate $2\lambda/\bar{X}.$ from the gamma distribution and, unlike the maximum likelihood estimate for ψ^{-1} in the gamma distribution, $\bar{X}./2\lambda$, we have that $\hat{\psi}_\lambda^{-1}$ is a biased estimate for ψ^{-1}.

Using (4.36), (4.37), (2.4) and (2.5) we have

$$\hat{\psi}_\lambda | \hat{\omega}_\lambda = w \sim N^{-1}(-n\lambda, n\psi(\lambda + \sqrt{\lambda^2 + w^2\bar{X}.\bar{X}_{-1}}), 0), \tag{4.38}$$

and (4.29) now follows using (4.2).

We now know the asymptotic distribution of $\hat{\omega}_\lambda$ and the conditional distribution of $\hat{\psi}_\lambda$ given $\hat{\omega}_\lambda$, and hence we have only to integrate out w in (4.38).

If we take a first order approximation to the square root in (4.37) we have

$$\hat{\psi}_\lambda \simeq \bar{X}_{.}^{-1} (2\lambda + \max(0, Z)),$$

where $Z = V_0\bar{X}.\bar{X}_{-1}$, and using (4.35) we have

$$\sqrt{n}\, Z \overset{\sim}{\rightarrow} N(0, 2\lambda(\lambda-2)) \quad (n \rightarrow \infty). \tag{4.39}$$

Let φ denote the approximate normal density for $2\lambda + Z$, according to (4.39), and let

$$g(x|y) = \frac{(y\frac{n\psi}{2})^{n\lambda}}{\Gamma(n\lambda)} x^{-n\lambda-1} e^{-\frac{yn\psi}{2x}}$$

be the density for the variate $y\bar{X}_{.}^{-1} \sim N^{-1}(-n\lambda, n\psi y, 0)$. For large n we then get the following approximate density for $\hat{\psi}_\lambda$ by integrating out w in (4.38), taking a first order approximation to the square root in (4.38),

$$f_{\hat{\psi}_\lambda}(x) = \frac{1}{2} g(x|2-\lambda) + \int_{2\lambda}^{\infty} g(x|y)\varphi(y)dy. \tag{4.40}$$

The integral in (4.40) is difficult to handle, but since the coefficient of variation of $Z + 2\lambda$ tends to zero as n tends to infinity we may approximate φ by a gamma density. By equating mean and variance of $Z+2\lambda$ to the mean and variance of the gamma distribution we have approximately for large n

$$Z+2\lambda \sim N^{-1}\left(\frac{2\lambda n}{\lambda-2}, 0, \frac{2n}{\lambda-2}\right),$$

and letting now φ be the corresponding density the integral in (4.40) may be expressed in terms of an incomplete gamma function. From this (4.30) follows by elementary calculations, and we omit the details.

From (2.18) we have for $k < n\lambda$

$$\int_0^{\infty} x^k g(x|y)dx = \frac{\Gamma(n\lambda-k)}{\Gamma(n\lambda)} \left(\frac{n\psi y}{2}\right)^k,$$

and using this and (4.40) we get (4.31). The results (4.32) and (4.33) are special cases of (4.31). **

4.3. The partially maximized log-likelihood for λ, estimation of λ.

This section concerns estimation in the full family of generalized inverse Gaussian distributions. The main task is of course to estimate λ since we can estimate (χ,ψ) for any given λ.

The principal tool for the estimation of λ is the partially maximized log-likelihood

$$\tilde{l}(\lambda) = \sup_{(\chi,\psi)\in\Theta_\lambda} l(\lambda,\chi,\psi)$$

which may be obtained by inserting the estimate $(\hat{\chi}_\lambda,\hat{\psi}_\lambda)$ in the log-likelihood (4.17). If we assume that not all observations are equal the log-likelihood function is strictly convace and has a unique maximum. These properties are preserved when we maximize out the component (χ,ψ) of the parameter , and hence $\tilde{l}$ is strictly concave and has a unique maximum which corresponds to the maximum likelihood estimate $\hat{\lambda}$. Having implemented the estimation of (χ,ψ) for fixed λ on the computer (according to Theorem 4.1), it is straight forward also to determine $\hat{\lambda}$ numerically, e.g. by a simple tabulation of $\tilde{l}$.

Let us denote the overall estimate of the parameter by $(\hat{\lambda},\hat{\chi},\hat{\psi})$, i.e. we have $(\hat{\chi},\hat{\psi}) = (\hat{\chi}_{\hat{\lambda}},\hat{\psi}_{\hat{\lambda}})$.

In general the Bessel function, and hence the likelihood, is difficult to handle as a function of the index λ, and analytical results about $\hat{\lambda}$ are difficult to obtain. Therefore we have only been able to obtain some approximate results concerning extreme properties of the partially maximized log-likelihood.

Using (4.1) we have

$$\tilde{l}(\lambda) =$$

$$n[-\lambda \ln \hat{\eta}_\lambda - \ln 2K_\lambda(\hat{\omega}_\lambda) + (\lambda-1)\bar{x}_\sim - \tfrac{1}{2}\hat{\omega}_\lambda (R_{-\lambda}(\hat{\omega}_\lambda) + R_\lambda(\hat{\omega}_\lambda))]$$

$$(|\lambda| < u), \qquad (4.41)$$

where $u = \bar{x}.\bar{x}_{\dashv}/(\bar{x}.\bar{x}_{\dashv} - 1)$. From (4.6) we have

$$\tilde{l}(\lambda) = \begin{cases} n[\lambda \ln \frac{\lambda}{\bar{x}.} - \ln\Gamma(\lambda) + (\lambda-1)\bar{x}_\sim - \lambda] & \text{if } \lambda \geq u \\ n[-\lambda \ln \frac{-\lambda}{\bar{x}_{\dashv}} - \ln\Gamma(-\lambda) + (\lambda-1)\bar{x}_\sim + \lambda] & \text{if } \lambda \leq -u . \end{cases} \qquad (4.42)$$

Let us first consider the case where $\bar{x}.\bar{x}_{\dashv}$ is close to 1 so that u and $\hat{\omega}_\lambda$ are large. Using (4.3), (4.11), (A.9) and (A.21) and taking a first order approximation to the logarithm we obtain the following asymptotic expansion of $\tilde{l}$ for $\hat{\omega}_\lambda$ large

$$\tilde{l}(\lambda) \simeq n[-\lambda(\ln\sqrt{\frac{\bar{x}.}{\bar{x}_{\dashv}}} - \frac{\lambda}{\bar{\omega}}) - (\ln(\sqrt{2\pi}\,\bar{\omega}^{-\frac{1}{2}} e^{-\bar{\omega}}) + \frac{\frac{1}{2}\lambda^2 - \frac{1}{8}}{\bar{\omega}})$$

$$+ (\lambda-1)\bar{x}_\sim - \tfrac{1}{2}\bar{\omega}\,(2 + \frac{1}{\bar{\omega}} + \frac{\lambda^2 - \frac{1}{4}}{\bar{\omega}^2})] \qquad (|\lambda| < u), \qquad (4.43)$$

where $\bar{\omega} = 1/(\bar{x}.\bar{x}_{\dashv} - 1)$. The coefficient of λ^2 is zero, and hence (4.43) is linear with slope $n[\bar{x}_\sim - \ln\sqrt{\frac{\bar{x}.}{\bar{x}_{\dashv}}}]$. When $\bar{x}.\bar{x}_{\dashv}$ is close to 1, $\tilde{l}$ will in practice have some curvature since the error using (4.43) for $\tilde{l}$ increases with $|\lambda|$.

The form of the tails of $\tilde{l}$ may be described as follows: If we approximate $\Gamma(\lambda)$ by Stirling's formula we have for $\lambda > u$, using (4.42),

$$\tilde{l}(\lambda) \simeq n[-\lambda \ln \bar{x}. + \frac{1}{2}\ln \lambda - \ln \sqrt{2\pi} + (\lambda-1)\bar{x}_{\sim}],$$

i.e. a linear and a logarithmic term. Since the curvature of the logarithm is small for large arguments the tails of $\tilde{l}$ will be almost linear.

Figure 4.3 shows an example of a partially maximized log-likelihood ($n = 26$, $\bar{x}.\bar{x}_{\dashv} = 1.089$, $u = 12.29$), for which both the above characteristics are prominent. We have taken a rather large range of values for λ in order to show the tails of $\tilde{l}$. Let us note that for this example we have $\hat{\lambda} = 12.5 > u$, and hence $\hat{\omega} = 0$.

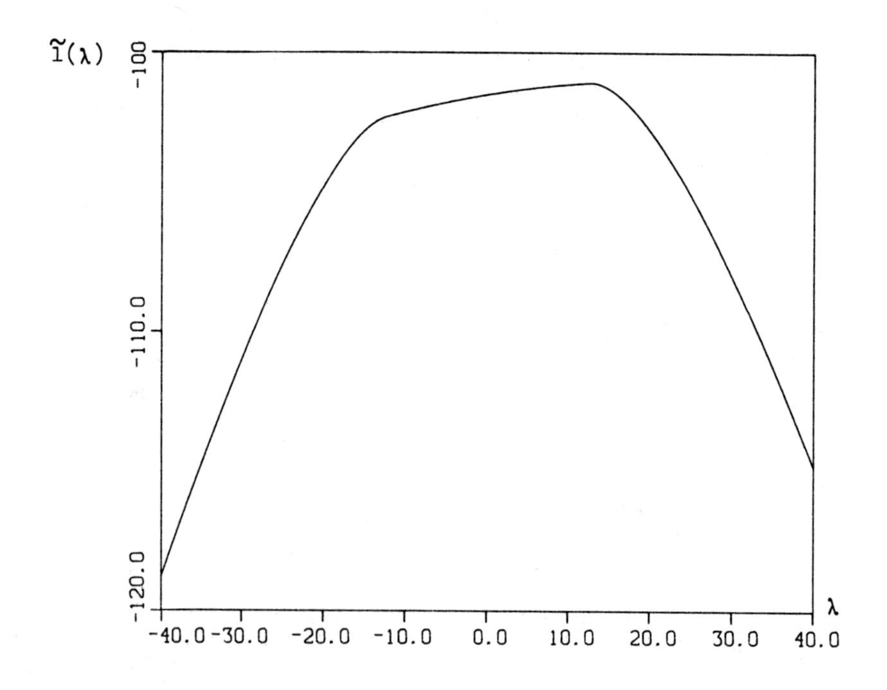

Figure 4.3. An extreme example of a partially maximized log-likelihood, where $\bar{x}.\bar{x}_{\dashv} = 1.089$ and $u = 12.29$.

Finally, let us consider the right derivative of $\tilde{l}$ at u. A quite similar discussion will apply for the left derivative at -u. These derivatives are interesting because they determine whether $\hat{\lambda}$ is inside or outside the interval $[-u,u]$ and hence whether $(\hat{\chi},\hat{\psi})$ belongs to $\text{int}\,\Theta_{\hat{\lambda}}$ or not.

The right derivative of $\tilde{l}$ at u is

$$\tilde{l}'(u^+) = n[\ln u \frac{\tilde{x}}{\bar{x}.} - \psi(u)], \qquad (4.44)$$

where $\tilde{x} = \exp \bar{x}_{\sim}$ is the geometric mean and where ψ is the digamma function. From (4.4) we have that for any value of u the variable $\tilde{x}/\bar{x}.$ may vary between $(u-1)/u$ and 1, and hence $\tilde{l}'(u^+)$ may vary in the interval

$$(n[\ln(u-1) - \psi(u)], n[\ln u - \psi(u)]), \qquad (4.45)$$

where the right endpoint is always positive and decreases towards zero as u becomes large. Hence we may have $\hat{\omega} = 0$ for any $u > 1$. The left endpoint of (4.45) tends to $-\infty$ at $u = 1$ and tends to 0 for u large; probably it is negative for all u, but we have not been able to show this. If this conjecture is true we have that $\tilde{l}'(u^+)$ may be either negative or positive for any $u > 1$.

We conclude that when u is large both $\tilde{l}'(u^+)$ and $\tilde{l}'(-u^-)$ will be close to zero, and hence $\tilde{l}$ will be flat over the range $-u < \lambda < u$. In the opposite case, when the spread of the observations is large, u is close to 1 and $\tilde{x}/\bar{x}.$ is much less than 1, and hence $\tilde{l}'(u^+) < 0$ and $\tilde{l}'(-u^-) > 0$, so that $|\hat{\lambda}| < u$. This completes our discussion of the partially maximized log-likelihood.

From standard asymptotic theory for exponential families we have for $\omega > 0$

$$\sqrt{n}[(\hat{\lambda},\hat{\chi},\hat{\psi}) - (\lambda,\chi,\psi)] \overset{\sim}{\to} N(0, i(\lambda,\chi,\psi)^{-1}) \quad (n \to \infty) \tag{4.46}$$

where i is the Fisher information matrix for a single observation,

$$i(\lambda,\chi,\psi) =$$

$$\begin{Bmatrix} V_{\lambda,\chi,\psi}(\ln X) & -\frac{1}{2}E_{\lambda-1,\chi,\psi}(\ln X) & -\frac{1}{2}E_{\lambda+1,\chi,\psi}(\ln X) \\ & \frac{1}{4}R_{-\lambda}(\omega)^2\eta^{-2}(D_{\lambda-1}(\omega)-1) & \frac{1}{4}(1 - D_{\lambda}(\omega)) \\ & & \frac{1}{4}R_{\lambda}(\omega)^2\eta^2(D_{\lambda+1}(\omega)-1) \end{Bmatrix}, \tag{4.47}$$

and where

$$E_{\lambda,\chi,\psi}(\ln X) = \int_0^\infty \frac{\eta^{-\lambda}}{2K_\lambda(\omega)} x^{\lambda-1} e^{-\frac{1}{2}(\chi x^{-1}+\psi x)} \ln x \, dx,$$

etc. In section 3.1 we derived approximations to the mean and variance of $\ln X$.

4.4. Estimation of ω when λ and η are fixed

Consider the full exponential family defined by the density function (2.2) where λ and η are fixed. This family is regular since ω varies in $(0,\infty)$ and hence the maximum likelihood estimate $\hat{\omega}_{\lambda,\eta}$ is the unique solution to the likelihood equation

$$\eta\bar{x}_{-1} + \eta^{-1}\bar{x}. = R_{-\lambda}(\omega) + R_{\lambda}(\omega).$$

In particular the function $f(\omega) = R_{-\lambda}(\omega) + R_{\lambda}(\omega)$ is monotone, in fact (A.21) shows that f is decreasing and maps $(0,\infty)$ onto $(2,\infty)$.

The Fisher information for a single observation is

$$i_{\lambda,\eta}(\omega) = \frac{1}{4}[R_{-\lambda}^{2}(\omega)(D_{\lambda-1}(\omega)-1)+R_{\lambda}^{2}(\omega)(D_{\lambda+1}(\omega)-1)+2(1-D_{\lambda}(\omega))].$$

4.5. Estimation of χ when λ and $\psi > 0$ are fixed

For fixed λ and $\psi > 0$ we have a full exponential family with canonical parameter χ and corresponding canonical statistic $-\frac{1}{2}x_{\dashv}$, the family being regular for $\lambda \leq 0$ and non-regular for $\lambda > 0$. If we follow the notation of the proof of Theorem 4.1 the mean value mapping is $\tau_{\lambda 1}(\cdot,\psi)$, where

$$\tau_{\lambda 1}(\chi,\psi) = -\frac{n}{2} R_{-\lambda}(\sqrt{\chi\psi})\sqrt{\psi/\chi}.$$

From the proof of Theorem 4.1, in particular (4.20), we have that the cumulant transform $\kappa_{\lambda}(\cdot,\psi)$ is steep when $\lambda \leq 1$ and not steep when $\lambda > 1$. From (A.21) we get

$$\lim_{\chi\to\infty} \tau_{\lambda 1}(\chi,\psi) = 0,$$

and hence $\tau_{\lambda 1}(\cdot,\psi)$, being a mean value mapping, is strictly increasing. In particular the function $f(\omega) = R_{\lambda}(\omega)/\omega$ is strictly decreasing.

The likelihood equation takes the form

$$R_{-\lambda}(\sqrt{\chi\psi})\sqrt{\psi/\chi} = \bar{x}_{\dashv}, \tag{4.48}$$

and for $\lambda \leq 1$ the maximum likelihood estimate $\hat{\chi}_{\lambda,\psi}$ is the unique solution to (4.48).

When $\lambda > 1$ the likelihood equation has by (4.20) a solution $\hat{\chi}_{\lambda,\psi} > 0$ if and only if

$$\bar{x}_{\dashv} < \frac{\psi}{2(\lambda-1)}.$$

In the opposite case the likelihood is decreasing and hence $\hat{\chi}_{\lambda,\psi} = 0$.

If we use (2.4) and consider the reciprocal observations the above results apply for the estimation of ψ when λ and $\chi > 0$ are fixed.

5. Inference

In this chapter we consider inference about the parameters of the generalized inverse Gaussian distribution.

Section 5.1 contains some exact and approximate distribution results.

In the two subsequent sections we consider inference about λ and ω when one has a single random sample from the generalized inverse Gaussian distribution. Section 5.2 concerns the inference about λ when χ and ψ are incidental and section 5.3 concerns inference about ω when λ is fixed and the scale parameter is incidental.

In section 5.4 we consider a one-way analysis of variance model, and in section 5.5 we consider a regression model.

5.1. Distribution results

In this and the following two sections we assume that $X_1,\dots,X_n$ are i.i.d. random variables with distribution $N^{-}(\lambda,\chi,\psi)$.

Let us consider the decomposition of the resultant vector into hyperbolic coordinates:

$$(X_{-},X.) = T(S^{-1},S),$$

where the variables

$$T = \sqrt{X.X_{-}}\ , \qquad S = \sqrt{X./X_{-}}\ ,$$

are called respectively the (hyperbolic) resultant length and the direction (of the resultant). This terminology, which stresses the analogy to the von Mises distribution, was partly introduced by Barndorff-Nielsen (1978b) in connection with the hyperbola distribution. In this section we shall mainly be concerned with the distribution of S and T.

The following two theorems are central for the inference, as will become clear in the later sections.

Theorem 5.1. Let $U_i = X_i/S$, $i = 1,\dots,n$. Then S and the vector $(U_1,\dots,U_n)$ are conditionally independent* given $T = t$ and

$$S|T = t \sim N^{-}(n\lambda, t\chi, t\psi). \tag{5.1}$$

The joint density of $U_1,\dots,U_{n-2}$, T is

---

* Note added in proof: Hall, Wijsman and Ghosh (Ann.Statist. 36 (1965), 575-614) have a general result about conditional independence in transformation families which gives the conditional independence in the present case. See also Barndorff-Nielsen, Blæsild, Jensen, and Jørgensen: "Exponential transformation models" (to appear in Proc.Roy.Soc.Ser.A).

$$\frac{K_{n\lambda}(t\omega)}{2^{n-3}K_\lambda(\omega)^n} t \prod_{i=1}^{n-2} u_i^{\lambda-1}\left(\frac{a'_\cdot}{a'_\dashv}\right)^\lambda (a'_\dashv a'_\cdot(a'_\dashv a'_\cdot - 4))^{-\frac{1}{2}}, \quad (a'_\dashv\, a'_\cdot > 4, t > n),$$

where $a'_\dashv = t - \sum_{i=1}^{n-2} u_i^{-1}$ and $a'_\cdot = t - \sum_{i=1}^{n-2} u_i$. ** (5.2)

<u>Theorem 5.2</u>. Assume that the statistic V is a function of S and T only, and let $U = u(X_1,\ldots,X_n)$ be any statistic which is invariant under scale transformations. Then U and V are conditionally independent given T, and the conditional distribution of $U|T = t$ depends on λ only. **

<u>Proof</u>. Let us first show that Theorem 5.2 is a consequence of Theorem 5.1.

Note that since U is invariant we may express U in terms of the U_i's from Theorem 5.1, viz.

$$U = u(X_1/S,\ldots,X_n/S).$$

Hence we have from Theorem 5.1 that, given $T = t$, U and S are conditionally independent and consequently that U and any function of t and S, $V = v(t,S)$, are conditionally independent. From the conditional independence of U and S we have that the conditional distributions of $U|T$ and $U|(S,T)$ are the same, but since (S,T) is sufficient for (χ,ψ) we have that the conditional distribution of $U|T$ depends only on λ.

For the proof of Theorem 5.1 we need the distribution of $(U_1,\ldots,U_{n-2},S,T,O)$, where $O = \mathrm{sgn}(\ln X_n/X_{n-1})$. First we transform by the mapping

$$f:(x_1,\ldots,x_n) \to (x_1,\ldots,x_{n-2},x_{\dashv},x_{\cdot},0).$$

Letting $a_{\dashv} = x_{\dashv} - \sum_{i=1}^{n-2} x_i^{-1}$ and $a_{\cdot} = x_{\cdot} - \sum_{i=1}^{n-2} x_i$, the inverse of f is determined by the equations

$$\mathrm{sgn}(\ln x_n/x_{n-1}) = 0,$$

$$x_{n-1}x_n = \frac{a_{\cdot}}{a_{\dashv}}$$

and

$$\frac{x_n}{x_{n-1}} + \frac{x_{n-1}}{x_n} + 2 = a_{\dashv}a_{\cdot} \ .$$

The solutions to the last equation are

$$\frac{x_n}{x_{n-1}} = \frac{a_{\dashv}a_{\cdot} - 2 \pm \sqrt{a_{\dashv}a_{\cdot}(a_{\dashv}a_{\cdot} - 4)}}{2}$$

and hence the norm of the Jacobian of f is

$$|x_n^{-2} - x_{n-1}^{-2}|$$

$$= \frac{\left|\frac{x_{n-1}}{x_n} - \frac{x_n}{x_{n-1}}\right|}{x_{n-1}x_n}$$

$$= \frac{a_{\dashv}}{a_{\cdot}}\sqrt{a_{\dashv}a_{\cdot}(a_{\dashv}a_{\cdot} - 4)}.$$

Thus the density of $(X_1,\ldots,X_{n-2},X_{\dashv},X_{\cdot},0)$ with respect to the product measure $\mu \times \tau$, where μ is the Lebesgue measure on $\mathbb{R}^n$ and τ is the counting measure, is

$$\frac{\eta^{-n\lambda}}{2^n K_\lambda(\omega)^n} \prod_{i=1}^{n-2} x_i^{\lambda-1} e^{-\frac{1}{2}(\chi x_{\dashv} + \psi x.)} \left(\frac{a.}{a_{\dashv}}\right)^{\lambda} (a_{\dashv}a.(a_{\dashv}a. - 4))^{-\frac{1}{2}},$$

$$(a_{\dashv}a. > 4,\ x_{\dashv}x. > n^2). \tag{5.3}$$

A change of variables in (5.3) to $(U_1,\ldots,U_{n-2},S,T,O)$ yields

$$\frac{\eta^{-n\lambda}}{2^{n-1} K_\lambda(\omega)^n} t \prod_{i=1}^{n-2} u_i^{\lambda-1} s^{n\lambda-1} e^{-\frac{1}{2}t(\chi s^{-1} + \psi s)} \left(\frac{a'.}{a'_{\dashv}}\right)^{\lambda} (a'_{\dashv}a'.(a'_{\dashv}a'. - 4))^{-\frac{1}{2}}$$

$$(a'_{\dashv}a'. > 4, t > n), \tag{5.4}$$

where

$$a'_{\dashv} = t - \sum_{i=1}^{n-2} u_i^{-1} \quad \text{and} \quad a'. = t - \sum_{i=1}^{n-2} u_i.$$

For any given t, (5.4) is the product of two factors, one involving only s, giving (5.1), and the other involving only $u_1,\ldots,u_{n-2}$ and o. Hence $(U_1,\ldots,U_n)$ and S are conditionally independent given $T = t$. Finally (5.2) follows by integrating out s and o in (5.4). **

Now let us make some comments on the results in Theorem 5.1.

The inference about the parameters λ, ω and η may be summarized in the following factorization of the density for the sufficient statistic (S,T,W):

$$p(w,s,t;\lambda,\chi,\psi) = p(t;\lambda,\omega)p(s|t;\lambda,\chi,\psi)p(w|t;\lambda), \tag{5.5}$$

where $W = \bar{X}_{\sim} - \ln S$. The factorization follows because W is invariant and hence, by Theorem 5.2, the statistics S and W are conditionally independent given T. The inference about λ and ω is discussed in detail in section 5.2 and 5.3.

Second, note that (5.1) is the conditional distribution of the direction given the resultant length, i.e. given that the resultant belongs to a hyperbola, and note that the generalized inverse Gaussian distribution in (5.1) has concentration parameter $t\omega$ and scale parameter as before. Hence the analogy to the von Mises distribution is rather close, particularly for the hyperbola distribution, $\lambda = 0$. In the latter case (5.1) has been shown independently by Rukhin (1974) and Barndorff-Nielsen (1978b).

Third, note that, from (2.5) and (5.1),

$$X.|T = t \sim N^{-}(n\lambda, t^2\chi, \psi),$$

since $X. = ST$. In particular this shows that for the gamma distribution the variables $X.$ and T are independent and $X. \sim N^{-}(n\lambda, 0, \psi)$. The independence was shown in section 4.2 using Basu's Theorem.

Finally, one consequence of (5.1) is that to find the distribution of the resultant we need only find the distribution of T. In principle we may integrate out $(u_1,\ldots,u_{n-2})$ in (5.2) to get the density of T, but in the general case this seems hardly feasible, either analytically, or numerically. Below we find the distribution of T in a few special cases, but first we consider the distribution of T from a different viewpoint. (We also consider the distribution of T in section 5.4.)

Let $h^n_{\lambda,\omega}(t)$ denote the density function for T and let $P_{\lambda,\chi,\psi}$ denote the probability measure associated with the distribution $N^{-}(\lambda,\chi,\psi)$. We then have for any fixed $(\chi_0,\psi_0) \in \Theta_\lambda$

$$\frac{dP^n_{\lambda,\chi,\psi}\cdot(X_{-},X.)}{dP^n_{\lambda,\chi_0,\psi_0}\cdot(X_{-},X.)}(x_{-},x.) = \left(\frac{\eta^{-\lambda}K_\lambda(\omega_0)}{\eta_0^{-\lambda}K_\lambda(\omega)}\right)^n e^{-\frac{1}{2}((\chi-\chi_0)x_{-}+(\psi-\psi_0)x.)}$$

and from (5.1)

$$\frac{dP^n_{\lambda,\chi_0,\psi_0}\cdot(S,T)}{d\mu^2}(s,t) = h^n_{\lambda,\omega_0}(t)\,\frac{\eta_0^{-n\lambda}}{2K_{n\lambda}(\omega_0 t)}\,s^{n\lambda-1}e^{-\frac{1}{2}t(\chi_0 s^{-1}+\psi_0 s)},$$

where μ denotes the Lebesgue measure, and hence

$$\frac{dP^n_{\lambda,\chi,\psi}\cdot(S,T)}{d\mu^2}(s,t) = h^n_{\lambda,\omega_0}(t)\left(\frac{\eta^{-\lambda}K_\lambda(\omega_0)}{K_\lambda(\omega)}\right)^n \frac{s^{n\lambda-1}}{2K_{n\lambda}(\omega_0 t)}e^{-\frac{1}{2}t(\chi s^{-1}+\psi s)}.$$

Thus we have, integrating out s,

$$h^n_{\lambda,\omega}(t) = h^n_{\lambda,\omega_0}(t)\,\frac{K_\lambda(\omega_0)^n K_{n\lambda}(t\omega)}{K_\lambda(\omega)^n K_{n\lambda}(t\omega_0)}. \tag{5.6}$$

Using (A.7) we can, for $\lambda \neq 0$, express $h^n_{\lambda,\omega}$ in terms of $h^n_{\lambda,0}$:

$$h^n_{\lambda,\omega}(t) = h^n_{\lambda,0}(t)\,\frac{\Gamma(|\lambda|)^n t^{n|\lambda|}K_{n\lambda}(t\omega)}{\Gamma(n|\lambda|)2^{n-1}\,K_\lambda(\omega)^n}. \tag{5.7}$$

From (5.7) and the fact that $h^n_{\lambda,\omega}$ depends on λ only through $|\lambda|$ it follows that it is sufficient to find the distribution of T for $\omega = 0$, $\lambda > 0$, i.e. for the gamma distribution; this, however, has not been possible.

For the hyperbola distribution, $\lambda = 0$, the density of T may be expressed in the following form (Rukhin, 1974):

$$h^n_{0,\omega}(t) = \frac{2K_0(\omega t)t}{K_0(\omega)^n\pi}\,\mathrm{Re}\left[i\int_0^\infty J_0(ts)K_0(is)^n s\,ds\right] \qquad (n\geq 3),$$

where J_0 is the standard Bessel function of order zero. One notes the similarity to the expression for the density of the resultant length in the von Mises distribution (Mardia (1972), p. 94-95).

Seshadri and Shuster (1974) showed that for $\lambda = -\frac{1}{2}$, $\psi = 0$ (and hence for $\lambda = \frac{1}{2}$, $\chi = 0$) we have

$$(T^2/n^2 - 1)/(n-1) \sim F_{n-1,1}. \tag{5.8}$$

The proof of (5.8) makes use of the fact that the distribution $N^{-}(-\frac{1}{2},\chi,0)$ is stable, and hence there seems to be no extension of the proof to general λ. Using (5.8), (5.7) and (5.1) it may be shown for the inverse Gaussian distribution that the variables $Y = X_{-} - n^2/X.$ and $X.$ are independent and

$$Y \sim N^{-}(\tfrac{n-1}{2},0,\chi), \quad X. \sim N^{-}(-\tfrac{1}{2},n^2\chi,\psi)$$

(Tweedie, 1957).

In the case $n = 2$ we have from (5.3) and (5.2)

$$\frac{dP^2_{\lambda,\chi,\psi}(Y,X.)}{d\mu^2}(y,x.) = \frac{\eta^{-2\lambda}}{2K_\lambda(\omega)^2} x.^{\lambda-1} e^{-\frac{1}{2}(4\chi x.^{-1} + \psi x.)}$$

$$\times \frac{e^{-\frac{1}{2}\chi y}}{(y + 4x.^{-1})^{\lambda+\frac{1}{2}}\sqrt{y}} \tag{5.9}$$

and

$$h^2_{\lambda,\omega}(t) = \frac{2K_{2\lambda}(t\omega)}{K_\lambda(\omega)^2}(t^2 - 4)^{-\frac{1}{2}}, \qquad (t > 2), \tag{5.10}$$

and in particular for $\omega = 0$

$$2|\lambda|(T^2/4-1) \sim F_{1,2|\lambda|}. \tag{5.11}$$

The results (5.11) and (5.8) seem to call for a generalization, but we have not been able to find it. The density (5.9) shows that in the case $n = 2$ the variables Y and $X.$ are independent only for $\lambda = -\frac{1}{2}$, and probably this is true for general $n \geq 2$.

We now turn to some approximate results.

From formula (3.12) of Barndorff-Nielsen and Cox (1979) we have the following saddle-point approximation to the joint density of $(X_{\dashv}, X.)$

$$q^n_{X_{\dashv},X.}(x_{\dashv}, x.;\lambda,\chi,\psi)$$

$$= \frac{\eta^{-n\lambda} K_\lambda(\hat{\omega}_\lambda)^n \hat{\eta}_\lambda^{n\lambda}}{K_\lambda(\omega)^n 2\pi n V_\lambda(\hat{\omega}_\lambda)^{\frac{1}{2}}} e^{-\frac{1}{2}(\chi x_{\dashv} + \psi x.) + \frac{1}{2}(\hat{\chi}_\lambda x_{\dashv} + \hat{\psi}_\lambda x.)}, \quad (x_{\dashv} x. > n^2). \quad (5$$

Here $V_\lambda(\omega)$ is the generalized variance of $\frac{1}{2}(X^{-1}, X)$,

$$V_\lambda(\omega) = \frac{1}{4}|i_\lambda(\chi,\psi)| = \frac{1}{16}[D_\lambda(\omega)^2(D_{\lambda-1}(\omega)-1)(D_{\lambda+1}(\omega)-1)-(D_\lambda(\omega)-1)^2],$$

and $(\hat{\chi}_\lambda, \hat{\psi}_\lambda)$ is the maximum likelihood estimate for fixed λ based on $(x_{\dashv}, x.)$, according to Theorem 4.1. If $\hat{\omega}_\lambda = 0$ (5.12) should be interpreted in the limiting sense.

If we change variables to (S,T) and reduce using (4.9) we have

$$q^n_{S,T}(s,t;\lambda,\chi,\psi)$$

$$= \frac{\eta^{-n\lambda} K_\lambda(\hat{\omega}_\lambda)^n \tilde{\eta}_\lambda^{n\lambda}}{K_\lambda(\omega)^n \pi n V_\lambda(\hat{\omega}_\lambda)^{\frac{1}{2}}} t\, s^{n\lambda-1} e^{-\frac{1}{2}t(\chi s^{-1}+\psi s)+\sqrt{(n\lambda)^2+\hat{\omega}_\lambda^2 t^2}}, \quad (t>n), \quad (5.$$

where

$$\tilde{\eta}_\lambda = \sqrt{\frac{R_{-\lambda}(\hat{\omega}_\lambda)}{R_\lambda(\hat{\omega}_\lambda)}}.$$

Somewhat surprisingly (5.13) yields again (5.1), except for the normalizing constant.

Integrating out s in (5.13) we have the following approximate density for T

$$q_T^n(t;\lambda,\omega)$$

$$= \frac{2K_\lambda(\hat{\omega}_\lambda)^n \tilde{\eta}^{n\lambda}}{K_\lambda(\omega)^n \pi n V_\lambda(\hat{\omega}_\lambda)} t K_{n\lambda}(t\omega) e^{\sqrt{(n\lambda)^2 + \hat{\omega}_\lambda t^2}}, \qquad (t > n). \qquad (5.14)$$

We note that (5.14) gives the same likelihood for ω as (5.7).

The approximate density (5.12) is the first term of an asymptotic expansion of the exact density of $(X_{\dashv}, X.)$, and the expansion will converge uniformly in $(\bar{x}_{\dashv}, \bar{x}.)$, provided $(\hat{\chi}_\chi, \hat{\psi}_\lambda)$ belongs to a given, but arbitrary, compact subset of int Θ_λ (see Barndorff-Nielsen and Cox, 1979). In the case $|\lambda| > 1$ this is a severe restriction on $(\bar{x}_{\dashv}, \bar{x}.)$ since $\hat{\omega}_\lambda = 0$ for $\bar{x}.\bar{x}_{\dashv} > |\lambda|/(|\lambda|-1)$. For example we saw in section 4.2 that for $|\lambda| > 2$, $\omega = 0$ and large n we have $P(\hat{\omega}_\lambda = 0) \simeq 0.5$. In the case $1 < |\lambda| \leq 2$ we have that $V_\lambda(0) = \infty$ and hence that (5.12) is zero for $\bar{x}.\bar{x}_{\dashv} > |\lambda|/(|\lambda|-1)$.

On the other hand the possibility that the saddle-point expansion could converge for such values of $(\bar{x}_{\dashv}, \bar{x}.)$ is not precluded beforehand, and Barndorff-Nielsen and Cox (1979) pointed out that (5.12) is exact for the inverse Gaussian distribution (and hence for $\lambda = \frac{1}{2}$), except for the normalizing constant. We return to this discussion in section 5.2, but we conclude here that a numerical investigation is called for.

Let us instead consider a large sample normal approximation.

The central limit theorem yields

$$\sqrt{n}[(\bar{X}_{-1},\bar{X}.)-(R_{-\lambda}(\omega)\eta^{-1},R_{\lambda}(\omega)\eta)] \overset{\sim}{\to} N(0,i_{\lambda}^{(1)}(\chi,\psi)) \qquad (5.15)$$

$$(n\to\infty),$$

where $i_{\lambda}^{(1)}(\chi,\psi)$ is the variance matrix (2.22) of (X^{-1},X). It is convenient to consider logarithms, whence

$$\sqrt{n}[(\ln\bar{X}_{-1},\ln\bar{X}.)-(\ln(R_{-\lambda}(\omega)\eta^{-1}),\ln(R_{\lambda}(\omega)\eta))] \overset{\sim}{\to} N(0,i_{\lambda}^{(2)}(\omega)) \qquad (5.16)$$

$$(n\to\infty),$$

where

$$i_{\lambda}^{(2)}(\omega) = \begin{Bmatrix} D_{\lambda-1}(\omega)-1 & D_{\lambda}(\omega)^{-1}-1 \\ D_{\lambda}(\omega)^{-1}-1 & D_{\lambda+1}(\omega)-1 \end{Bmatrix}$$

and

$$\sqrt{n}[(\ln S,\ln\tfrac{T}{n})-(\tfrac{1}{2}\ln(\frac{R_{\lambda}(\omega)}{R_{-\lambda}(\omega)}\eta^{2}),\tfrac{1}{2}\ln D_{\lambda}(\omega))] \overset{\sim}{\to} N(0,i_{\lambda}^{(3)}(\omega)) \qquad (5.17)$$

$$(n\to\infty),$$

where

$$i_{\lambda}^{(3)}(\omega) = \frac{1}{4}\begin{Bmatrix} D_{\lambda+1}(\omega)+D_{\lambda-1}(\omega)-2D_{\lambda}(\omega)^{-1} & D_{\lambda+1}(\omega)-D_{\lambda-1}(\omega) \\ D_{\lambda+1}(\omega)-D_{\lambda-1}(\omega) & D_{\lambda+1}(\omega)+D_{\lambda-1}(\omega)+2D_{\lambda}(\omega)^{-1}-4 \end{Bmatrix}$$

For $|\lambda|\leq 2$ the results (5.15), (5.16) and (5.17) are valid for $\omega>0$, whereas for $|\lambda|>2$ they are valid for $\omega\geq 0$.

5.2. Inference about λ

Inference about λ is, for several reasons, of primary interest. First, there are several values of λ that correspond to more or less well-known distributions: $\lambda = 0, \pm\frac{1}{2}, 1$, say, and, if possible, one often chooses one of these before carrying out a more detailed analysis. The reason for this is partly that inference in the families $N_\lambda^{\dashv}$ may have quite different structures for different values of λ, and partly a matter of convenience since the inference is much less complicated when λ is fixed. Furthermore the families $N_\lambda^{\dashv}$, having two parameters, are still quite flexible. Finally, the sign and the order of magnitude of λ is important, for example a negative sign of λ excludes the possibility of a gamma distribution.

For any fixed value of λ the statistic $(X_{\dashv}, X.)$ is sufficient and complete and hence inference about λ when χ and ψ are incidental should be carried out in the conditional distribution

$$X_{\sim} | X_{\dashv} = x_{\dashv}, \quad X. = x. \tag{5.18}$$

or equivalently in

$$\bar{X}_{\sim} - \ln S | T = t, \tag{5.19}$$

as $W = \bar{X}_{\sim} - \ln S$ is invariant and hence W and S conditionally independent given T (Theorem 5.2). Hence the inference about λ should be based on the factor $p(w|t;\lambda)$ in (5.5).

For fixed $\lambda_0 \in \mathbb{R}$ we have

$$\frac{dP_\lambda(W|T=t)}{dP_{\lambda_0}(W|T=t)}(w) = E_{\lambda_0}(e^{(\lambda-\lambda_0)W}|T=t)e^{(\lambda-\lambda_0)w}$$

an exponential family with W as canonical statistic and λ as canonical parameter, and hence the likelihood ratio is monotone in W for fixed value of t, so it is natural to take W as a test statistic. Instead of W we might of course consider an equivalent statistic such as for example $Q = \exp W = \tilde{X}\sqrt{X_{-1}/X.}$ (where $\tilde{X}$ is the geometric mean). the statistic Q having a certain intuitive appeal, but below we give some further arguments for preferring W.

The statistic W is a measure of the symmetry of the observations on a log-scale. In the untransformed scale this means that the order statistics of $U_1,\dots,U_n$ (in the notation of Theorem 5.1) are the same as the order statistics of $U_1^{-1},\dots,U_n^{-1}$ if and only if $W = 0$. The distribution of W is symmetric if and only if $\lambda = 0$ because the distribution of $\ln X$, where $X \sim N^{-}(\lambda,\chi,\psi)$, is symmetric if and only if $\lambda = 0$. Equation (4.43) and the discussion just after (4.43) indicate that a positive sign of W corresponds to a positive sign of λ and vice versa, and the expression (4.41) for the partially maximized log-likelihood $\tilde{l}$ contains the linear term $n\lambda(\bar{x}_{\sim} - \ln s)$.

Thus there are a number of arguments that lead us to draw the inference about λ in the conditional distribution of $W|T = t$ and to take W as a test statistic, large values of W indicating large values of λ and vice versa.

But unfortunately we do not know the exact distribution of $W|T = t$, so let us consider the saddle-point approximation.

The double saddle-point approximation to the density of the conditional distribution (5.18) is

$$q_{X_\sim | X_\dashv, X.}(x_\sim | x_\dashv, x.;\lambda)$$

$$= \frac{\left(\dfrac{\hat{\eta}^{-\hat{\lambda}} K_\lambda(\hat{\omega}_\lambda)}{K_{\hat{\lambda}}(\hat{\omega})\, \hat{\eta}_\lambda^{-\lambda}}\right)^n}{\sqrt{2\pi n}\,(V(\hat{\lambda},\hat{\chi},\hat{\psi})/V_\lambda(\hat{\omega}_\lambda))^{1/2}} e^{x_\sim(\lambda-\hat{\lambda}) - \frac{1}{2}((\hat{\chi}_\lambda-\hat{\chi})x_\dashv + (\hat{\psi}_\lambda-\hat{\psi})x.)}$$

$$(x_\dashv x. > n^2). \qquad (5.20)$$

Here V_λ is defined just after (5.12) and $V(\lambda,\chi,\psi)$ is the generalized variance of $(\ln X, \frac{1}{2}X^{-1}, \frac{1}{2}X)$. Subscript λ (e.g. $\hat{\omega}_\lambda$) denotes estimates for λ fixed whereas no subscript (e.g. $\hat{\omega}$) denotes overall estimates.

The approximation (5.20) is derived from saddle-point approximations to the distributions of $(X_\sim, X_\dashv, X.)$ and $(X_\dashv, X.)$, where we have already discussed the saddle-point approximation to the latter distribution in the foregoing section. In the present context it seems disasterous that the approximate density for $(X_\dashv, X.)$ may become zero. This becomes apparent if one considers the approximate conditional log-likelihood obtained from (5.20) by taking terms that depend on λ:

$$\frac{1}{2}\ln V_\lambda(\hat{\omega}_\lambda) + \tilde{l}(\lambda), \qquad (5.21)$$

where $\tilde{l}$ is the partially maximized log-likelihood for λ. In the case $1 \le u \le 2$ (where $u = \bar{x}.\bar{x}_\dashv/(\bar{x}.\bar{x}_\dashv - 1)$) (5.21) is infinite for $u \le |\lambda| \le 2$, and in some cases this even inflates (5.21) in the range $0.5 < |\lambda| < 1$. Figure 5.1 shows a severe case of this where $u = 1.04$ and $n = 9$. The figure shows $\tilde{l}$ and (5.21), where both functions have been normalized to have zero maximum for $\lambda \in [-1,1]$.

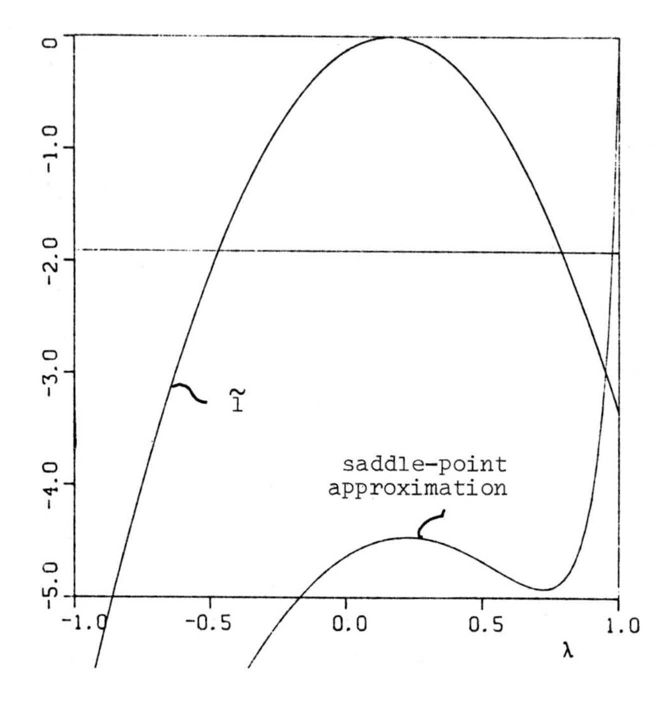

Figure 5.1. The partially maximized log-likelihood and the saddle-point approximation to the true conditional log-likelihood, for a sample with $n = 9$ and $u = 1.04$. Both functions have been normalized to have zero maximum. The horizontal line indicates the 95% confidence interval based on the asymptotic distribution ($\chi^2(1)$) of $-2\ln Q$, where Q is the likelihood ratio.

One notes that if $\hat{\omega}_\lambda = \hat{\omega} = 0$ the effect of V_λ being infinite is eliminated in (5.20) by V also being infinite.

At this point it should of course be stressed that the conditions for convergence of the saddle-point expansion are not fullfilled in the case where (5.21) breaks down, but as noted there is also an effect in the interval $-1 \leq \lambda \leq 1$ where the saddle-point expansion converges for all $(\bar{x}_{+}, \bar{x}.)$.

We have also tried to use the saddle-point approximation in some other situations concerning inference for the generalized inverse Gaussian distribution. One example where the approximation also breaks down was pointed out by Jørgensen and Pedersen (1979) in the discussion of Barndorff-Nielsen and Cox (1979). In that example we want to draw inference about ψ, λ being fixed and χ being incidental. It was shown by Jørgensen and Pedersen that in the case $1<\lambda\leq 2$ the saddle-point approximation to the conditional log-likelihood $l(\psi;x.|x_{-})$ is infinite in the interval $0<\psi\leq 2(\lambda-1)\bar{x}_{-}$.

In chapter 7 where we analyze some sets of data we shall use $\tilde{l}$ as a kind of approximation to the exact conditional log-likelihood for λ, having for the present no other choice. However, the degree of approximation involved when using $\tilde{l}$ in this way is not known.

We may also use $\tilde{l}$ to form the likelihood ratio test statistic for the hypothesis $\lambda = \lambda_0$, since

$$-2\ln Q = 2(\tilde{l}(\hat{\lambda}) - \tilde{l}(\lambda_0)),$$

where Q is the likelihood ratio, and if $(\chi,\psi)\in \operatorname{int}\Theta_{\lambda_0}$ we have $-2\ln Q \overset{\sim}{\to} \chi^2(1)$ under the null hypothesis. (Here $\chi^2(k)$ denotes the chi-squared distribution with k degrees of freedom.) Clearly one should use this result with care if $\hat{\omega} = 0$ or $\hat{\omega}_{\lambda_0} = 0$.

Another large-sample test concerning λ is obtained by referring $\hat{\lambda}$ to its asymptotic normal distribution obtained from (4.46).

5.3. Inference about ω

Let us now consider inference about ω when λ is known and the scale parameter incidental.

Recall from (5.5) that we have the following factorization of the density for the sufficient statistic (W,S,T):

$$p(w,s,t;\lambda,\chi,\psi) = p(t;\lambda,\omega)p(s|t;\lambda,\chi,\psi)p(w|t;\lambda),$$

where, for any fixed ω and t, the family $\{p(s|t;\lambda,\chi,\psi): \sqrt{\chi\psi} = \omega\}$ is universal and is generated by the group of scale transformations on $\mathbb{R}_+$. Hence T is both M- and G-sufficient for ω, in the sense of Barndorff-Nielsen (1978a), and we are led to draw the inference about ω in the marginal distribution of T. The distribution of T has already been considered in section 5.1.

One recalls here that T is a measure of the dispersion of the observations, whereas ω is a concentration parameter, and hence a small value of T indicates a large value of ω and vice versa.

Consider for a moment the situation where we want to test $\omega = 0$ or, equivalently, $\chi = 0$, for fixed $\lambda > 0$. Since χ is a canonical parameter in an exponential family we may test the hypothesis in the conditional distribution $X_{\dashv}|X. = x.$ or, equivalently, in the conditional distribution $T|X. = x.$. But, as noted earlier, T and $X.$ are independent in this case, and hence we are again led to test in the marginal distribution of T. A similar argument may be applied for $\lambda < 0$.

In the rest of this section we consider maximum likelihood

estimation of ω in the marginal distribution of T. Let us first consider the likelihood and its derivatives.

From (5.6) the log-likelihood becomes

$$l_\lambda(\omega) = \ln K_{n\lambda}(t\omega) - n \ln K_\lambda(\omega),$$

and using (A.9) we get the asymptotic relation

$$l_\lambda(\omega) \simeq -\omega(t-n) - \frac{1}{2}\ln t + \frac{n-1}{2}\ln\omega \qquad (\omega \to \infty),$$

i.e. a linear plus a logarithmic term. For $\lambda \neq 0$ we have by (A.7) that l_λ takes a finite value at $\omega = 0$,

$$l_\lambda(0) = \ln\left(\frac{\Gamma(n|\lambda|)}{\Gamma(|\lambda|)^n}\, 2^{n-1}\right) - n|\lambda|\ln t,$$

whereas for $\lambda = 0$ we have $l_0(\omega) \to -\infty$ as $\omega \downarrow 0$.

Using (A.3), (A.16) and (A.17) we get

$$l'_\lambda(\omega) = n\,R_\lambda(\omega) - t\,R_{n\lambda}(t\omega) \tag{5.22}$$

$$= n\sqrt{\left(\frac{\lambda}{\omega}\right)^2 + D_\lambda(\omega)} - t\sqrt{\left(\frac{n\lambda}{t\omega}\right)^2 + D_{n\lambda}(t\omega)}. \tag{5.23}$$

The expressions (5.22) and (5.23) give rise to two equivalent forms of the likelihood equation, namely

$$nR_\lambda(\omega) = t\,R_{n\lambda}(t\omega), \tag{5.24}$$

and

$$D^n_{\lambda,t}(\omega) = t^2/n^2, \tag{5.25}$$

where

$$D^n_{\lambda,t}(\omega) = \frac{D_\lambda(\omega)}{D_{n\lambda}(t\omega)}.$$

The second derivative of the log-likelihood is

$$l_\lambda''(\omega) = nR_\lambda^2(\omega) - t^2R_{n\lambda}(t\omega) - n\frac{2\lambda+1}{\omega}R_\lambda(\omega)$$
$$+ t\frac{2n\lambda+1}{\omega}R_{n\lambda}(t\omega) - n + t^2,$$

where we have used (5.22) and (A.23), and hence, using (5.24) and (A.16), we obtain the following expression for the observed information

$$\hat{j}_\lambda = (n^2 - n)D_\lambda(\hat{\hat{\omega}}_\lambda) - t^2 + n, \tag{5.26}$$

where $\hat{\hat{\omega}}_\lambda$ is a solution to the likelihood equation (5.24).

By similar manipulations we obtain the third derivative of l_λ at $\hat{\hat{\omega}}_\lambda$,

$$l_\lambda'''(\hat{\hat{\omega}}_\lambda) = 2(n - n^3)R_\lambda^3(\hat{\hat{\omega}}_\lambda) + 3n\frac{2\lambda(n^2-1)+n-1}{\hat{\hat{\omega}}_\lambda}R_\lambda^2(\hat{\hat{\omega}}_\lambda)$$
$$+ \left(\frac{4\lambda^2(n-n^3)+6\lambda(n-n^2)}{\hat{\hat{\omega}}_\lambda^2} + 2t^2n - 2n\right)R_\lambda(\hat{\hat{\omega}}_\lambda) + \frac{2\lambda n(1-t^2)+n-t^2}{\hat{\hat{\omega}}_\lambda}.$$

From this and (5.26) we may obtain an expression for Sprott's measure for the deviation of l_λ from normality,

$$|F_\omega| = |l_\lambda'''(\hat{\hat{\omega}}_\lambda)\hat{j}_\lambda^{-3/2}| \tag{5.27}$$

(Sprott, 1973). We need the expressions (5.26) and (5.27) in chapter 7 where we present some numerical examples. There we show, among other things, examples of graphs of the log-likelihood l_λ.

In order to prepare ourselves for the discussion of the

likelihood equation (5.25), let us consider the function $D^n_{\lambda,t}(\omega)$. From Theorem A.1 and the fact that $D_\lambda(\cdot)$ is greater than 1 and decreasing we have for $t > 1$ and $n \geq 1$

$$1 < D^n_{\lambda,t}(\omega) < D_\lambda(\omega),$$

and from (A.22)

$$D^n_{\lambda,t}(\omega) = 1 + \frac{1}{\omega}(1 - \frac{1}{t}) - \frac{1}{t}(1 - \frac{1}{t})\frac{1}{\omega^2} + O(\omega^{-3}) \qquad (\omega \to \infty). \qquad (5.28)$$

Furthermore we have from the asymptotic relations for D_λ listed in section 4.1 that

$$\lim_{\omega \downarrow 0} D^n_{\lambda,t}(\omega) = \begin{cases} \infty & \text{if } |\lambda| \leq 1 \\ \dfrac{|\lambda| - \frac{1}{n}}{|\lambda| - 1} & \text{if } |\lambda| > 1 \, . \end{cases}$$

These properties and the continuity of $D^n_{\lambda,t}$ indicate that $D^n_{\lambda,t}$ is decreasing, and plots of $D^n_{\lambda,t}$ for $\lambda = 0,\ 0.5,\ 1$ and 2 have shown no evidence against this assertion. Hence we make the following

<u>Conjecture 5.3</u>. The function $D^n_{\lambda,t}(\cdot)$ is decreasing for any $n \geq 2$, $\lambda \in \mathbb{R}$ and $t > n$. **

We can then prove

<u>Theorem 5.4</u>. Assume that Conjecture 5.3 is fullfilled. Let $\hat{\hat{\omega}}_\lambda$ denote the maximum likelihood estimate for ω in the marginal distribution of T. If $|\lambda| > 1$ and $\frac{t^2}{n^2} \geq \frac{|\lambda| - 1/n}{|\lambda| - 1}$ we have $\hat{\hat{\omega}}_\lambda = 0$. In the opposite case $\hat{\hat{\omega}}_\lambda$ is the unique solution to the likelihood equation (5.25), except if the observations

are all equal in which instance the likelihood does not attain its supremum.

Furthermore we have

$$\hat{\hat{\omega}}_\lambda \leq \hat{\omega}_\lambda \tag{5.29}$$

and for $\omega > 0$

$$\hat{\hat{\omega}}_\lambda = \hat{\omega}_\lambda + o_p(\tfrac{1}{n}) \qquad (n \to \infty) \tag{5.30}$$

$$\sqrt{n}(\hat{\omega}_\lambda - \omega) \overset{\sim}{\to} N(0, i_\lambda(\omega)^{-1}) \quad (n \to \infty) \tag{5.31}$$

$$\sqrt{n}(\hat{\hat{\omega}}_\lambda - \omega) \overset{\sim}{\to} N(0, i_\lambda(\omega)^{-1}) \quad (n \to \infty), \tag{5.32}$$

where

$$i_\lambda(\omega)^{-1} = \frac{R_\lambda(\omega)^2(D_{\lambda+1}(\omega)-1)+2(D_\lambda(\omega)-1)+R_{-\lambda}(\omega)^2(D_{\lambda-1}(\omega)-1)}{D_\lambda(\omega)^2(D_{\lambda-1}(\omega)-1)(D_{\lambda+1}(\omega)-1)-(D_\lambda(\omega)-1)^2} .$$

**

<u>Remark</u>. It is an open question whether the observed information $\hat{j}_\lambda$ (cf. (5.26)) gives, at least asymptotically, the precision of the maximum likelihood estimate $\hat{\hat{\omega}}_\lambda$. One might for example conjecture that, in some sense, the following asymptotic relation holds

$$\hat{j}_\lambda/n \to i_\lambda(\omega) \qquad (n \to \infty).$$

Some calculations in connection with one of the examples that we consider in chapter 7 do in fact indicate that this conjecture is correct. **

Proof. The first part of the theorem, including (5.29), is an immediate consequence of Conjecture 5.3 and the properties of $D^n_{\lambda,t}$.

From (2.19), (2.20), (A.14) and the strong law of large numbers we have

$$T^2/n^2 \to D_\lambda(\omega) \quad \text{a.s.} \quad (n \to \infty). \tag{5.33}$$

Now suppose that $\omega > 0$ and let us prove (5.30). First, note that $\hat{\hat{\omega}}_\lambda$ increases with n for given value of T/n (follows from the monotonicity properties of $D_\lambda(\omega)$). Hence, using (5.33) and (5.25) it follows that $\hat{\hat{\omega}}_\lambda^{-1} = O_p(1)$ $(n \to \infty)$.

Using (5.25) and (A.22) we get (as $n \to \infty$)

$$\begin{aligned} D_\lambda(\hat{\hat{\omega}}_\lambda)^{\frac{1}{2}} &= \frac{T}{n} D_{n\lambda}(T\,\hat{\hat{\omega}}_\lambda)^{\frac{1}{2}} \\ &= \frac{T}{n}\left(1 + O_p\left(\frac{1}{T}\right)\right) \\ &= \frac{T}{n} + O_p\left(\frac{1}{n}\right). \end{aligned}$$

Expanding $(D_\lambda^{1/2})^{-1}$, the function inverse of $D_\lambda^{1/2}$, around T/n we thus obtain, using (4.2),

$$\hat{\hat{\omega}}_\lambda = (D_\lambda^{\frac{1}{2}})^{-1}\left(\frac{T}{n} + O_p\left(\frac{1}{n}\right)\right) = \hat{\omega}_\lambda + O_p\left(\frac{1}{n}\right),$$

which proves (5.30).

From (4.25) we get (5.31), and (5.32) is a consequence of (5.31) and (5.30). **

The results in Theorem 5.4 are in fairly close analogy to the results of Schou (1978), who considered inference about the concentration parameter in the von Mises distribution. The results in Schou's paper and the similarity between the different kinds of Bessel functions suggest that it should be possible to obtain a proof of Conjecture 5.3 by standard methods.

In the case $\lambda = -\frac{1}{2}$ we are able to prove completely that the marginal likelihood has a unique maximum. This follows because $R_{-n/2}$ is an increasing function which maps $(0,\infty)$ onto $(0,1)$, whence the likelihood equation (5.24) has, by (A.18), a unique solution corresponding to a maximum for the likelihood. As we have $l_\lambda = l_{-\lambda}$ the same result is obtained for $\lambda = \frac{1}{2}$.

Having established Theorem 5.4 (supposing Conjecture 5.3 to be true) we are fairly well off concerning the estimation of ω, and in our experience the marginal likelihood equation (5.25) is just as easy to solve numerically as the equation $D_\lambda(\omega) = \bar{x}.\bar{x}_{-}$ which we considered in chapter 4.

To test hypotheses about ω we need the distribution of T. This is known exactly for $\lambda = \pm\frac{1}{2}$ (cf. (5.7) and (5.8)). In the cases $\omega > 0$, $\lambda \in \mathbb{R}$ and $\omega = 0$, $|\lambda| > 2$ the asymptotic distribution of T is provided by formula (5.17). For $\omega > 0$ and any λ we may also test in the asymptotic distribution of $\hat{\hat{\omega}}_\lambda$ which is given by (5.32).

The test for $\omega = 0$ or, equivalently, $\psi = 0$ in the inverse Gaussian distribution based on (5.8) was proposed by Nádas (1973) and has also been considered by Seshadri and Shuster (1974). As emphasized by these authors this is a test for zero drift in Brownian motion based on first passage times (cf. chapter 6).

5.4 One way analysis of variance

Suppose that we have k samples

$$X_{i1},\ldots,X_{in_i}, \quad i = 1,\ldots,k$$

of independent generalized inverse Gaussian variables, where

$$X_{ij} \sim N^{-}(\lambda,\chi_i,\psi_i), \quad i = 1,\ldots,k, \quad j = 1,\ldots,n_i,$$

and suppose that inference is wanted about the parameters in this model, λ being fixed.

For the inverse Gaussian distribution, assuming that there is a common value for χ, there is a complete analogue to the one-way analysis of variance model in the normal distribution, see Chhikara and Folks (1978). In that model η plays the role of the mean and χ^{-1} plays the role of the variance. However, there seems to be no scope for a generalization of the model to general λ because the simplicity of the model hinges on the similarity between the distributional theory for the inverse Gaussian distribution and the normal distribution.

Let us instead assume, inspired by the analogy to the von Mises distribution, that there is a common value for the concentration parameter,

$$\omega_i = \omega, \quad i = 1,\ldots,k.$$

Note that when ω and λ are fixed the only difference between the k distributions is now a scale parameter.

Suppose we want to test the hypothesis H_0 that the k distributions are identical, the alternatives being that at least two of the distributions are different. We shall now consider a test for this hypothesis which parallels the test for equal directions in the von Mises distribution (cf. Mardia, 1972).

First we need some notation. Let

$$X_{i\dashv} = \sum_{j=1}^{n_i} X_{ij}^{-1}$$

$$X_{i.} = \sum_{j=1}^{n_i} X_{ij}$$

$$T_i = \sqrt{X_{i.}X_{i\dashv}}$$

$$S_i = \sqrt{X_{i.}/X_{i\dashv}}$$

$$T = \sqrt{\sum_{i=1}^{k} X_{i.} \sum_{i=1}^{k} X_{i\dashv}}\ ,$$

i.e. T_i denotes the resultant length for the i'th sample and T denotes the resultant length for the total sample.

It now follows from Theorem 5.2 that under H_0 the conditional distribution

$$T_1,\ldots,T_k|T=t \tag{5.34}$$

depends on λ only, since $T_1,\ldots,T_k$ is invariant. Hence, for fixed λ we may test the hypothesis H_0 in the conditional distribution (5.34).

Under H_0 we expect $T_1+\ldots+T_k$ to be near T, whereas under the alternative hypothesis we expect T to be large compared with $T_1+\ldots+T_k$. Hence we suggest testing H_0 in the conditional distribution of $T_1+\ldots+T_k|T=t$, small values of

$T_1 + \ldots + T_k$ being significant. Clearly this is an analogy to the analysis of variance for the normal distribution. Note that Theorem 5.2 shows that under H_0 the test statistic and the estimate of the scale parameter are conditionally independent given T. This may perhaps be said to be the analogy to the independence of the estimator and the test statistic in the normal analysis of variance model.

We now consider the distribution of the test statistic, and later consider estimation in the model.

First we derive an inequality concerning the resultant length. With notation as above one has for $k = 2$

$$\begin{aligned} T &= \sqrt{(X_{1.} + X_{2.})(X_{1\dashv} + X_{2\dashv})} \\ &= \sqrt{(T_1 S_1 + T_2 S_2)(T_1/S_1 + T_2/S_2)} \\ &= \sqrt{T_1^2 + T_2^2 + T_1 T_2 (S_1/S_2 + S_2/S_1)}. \end{aligned} \tag{5.35}$$

For any sample one has $S_1/S_2 + S_2/S_1 \geq 2$, and hence we have the inequality

$$T \geq T_1 + T_2,$$

with equality if and only if $S_1 = S_2$. By a simple induction argument one has for general k

$$T \geq T_1 + \ldots + T_k, \tag{5.36}$$

with equality if and only if $S_1 = \ldots = S_k$.

Before proceeding to the distributional results we give an interpretation of (5.36). We shall see that the resultant length is a measure of the information in the sample and plays the role of sample size. Hence an interpretation of (5.36) is that the sum of the informations in the k subsamples is smaller than the information in the total sample.

To illustrate this, consider for example the situation where λ and ω are known (let us for simplicity assume $\lambda = 0$). Then inference about η (from a single sample) should be drawn in the conditional distribution of the maximum likelihood estimator $\hat{\eta} = S$ given the ancillary T, which, by (5.1), is a hyperbola distribution with concentration parameter $t\omega$. Formula (5.33) shows that $T/n \to D_\lambda(\omega)^{1/2}$ almost surely and hence formula (3.11) indicates that for large n the conditional distribution of $\ln \hat{\eta}$ given $T = t$ is approximately normal with mean $\ln \eta$ and variance $1/t\omega$. Hence the resultant length enters as a factor in the precision of the maximum likelihood estimate in exactly the same way as the sample size usually does. In the present case (5.36) shows that there is a loss of efficiency under H_0 by combining estimates of $\ln \eta$ from the k subsamples instead of considering the total sample.

After this short digression we return to the distribution of the test statistic. Let us first consider the conditional distribution

$$T|T_1 = t_1,\ldots,T_k = t_k. \tag{5.37}$$

Here and in the following we consider only distributions under H_0. First note that if we know the distribution of (5.37) and the distribution of the resultant length then the distribution (5.34) is easily derived, and from (5.34) the distribution of $T_1 + \ldots + T_k$ given T follows by integration.

In general it is not simple to find the distribution (5.37), and here we consider only the case $k = 2$. From (5.35) it follows that in order to find the conditional distribution of T given

T_1 and T_2 we must find the conditional distribution of S_2/S_1 given T_1 and T_2. Formula (5.1) gives

$$S_i | T_i = t_i \sim N^{-}(n_i\lambda, t_i\chi, t_i\psi), \quad i = 1,2 ,$$

and hence, the two samples being independent, we may use formula (3.20) which gives the density for a quotient between two independent generalized inverse Gaussian variates. By inserting the parameter values $(n_i\lambda, t_i\chi, t_i\psi)$ in (3.20) we obtain the following expression for the conditional density of $U = S_2/S_1$ given $T_1 = t_1$, and $T_2 = t_2$

$$\frac{K_{n.\lambda}(\omega\sqrt{t_1t_2(u+u^{-1}) + t_1^2 + t_2^2})}{2K_{n_1\lambda}(t_1\omega)K_{n_2\lambda}(t_2\omega)} \left(\frac{u^{1/2}t_1 + u^{-1/2}t_2}{u^{-1/2}t_1 + u^{1/2}t_2}\right)^{\frac{n.\lambda}{2}} u^{\frac{(n_2-n_1)\lambda}{2} - 1}$$

$$(t_1 > n_1,\ t_2 > n_2,\ u > 2), \qquad (5.38)$$

where $n. = n_1 + n_2$. Using (5.35) we can then transform (5.38) into the conditional density of T given T_1 and T_2

$$f(t|t_1,t_2;\lambda,\omega,n_1,n_2)$$

$$= \frac{K_{n.\lambda}(t\omega)t\left[\left(\frac{u^{1/2}t_1+u^{-1/2}t_2}{u^{-1/2}t_1+u^{1/2}t_2}\right)^{\frac{n.\lambda}{2}} u^{\frac{(n_2-n_1)\lambda}{2}} + \left(\frac{u^{1/2}t_1+u^{-1/2}t_2}{u^{-1/2}t_1+u^{1/2}t_2}\right)^{-\frac{n.\lambda}{2}} u^{-\frac{(n_2-n_1)\lambda}{2}}\right]}{K_{n_1\lambda}(t_1\omega)K_{n_2\lambda}(t_2\omega)\sqrt{(t^2-t_1^2-t_2^2)^2 - 4t_1^2t_2^2}}$$

$$(t_1 > n_1,\ t_2 > n_2,\ t > t_1 + t_2) \qquad (5.39)$$

where now $u = \dfrac{t^2 - t_1^2 - t_2^2 + \sqrt{(t^2 - t_1^2 - t_2^2)^2 - 4t_1^2t_2^2}}{2t_1t_2}$.

We may view (5.39) as giving a relation between the distributions of the resultant length for different sample sizes. In fact we have the formula

$$h^{n.}_{\lambda,\omega}(t) = \iint_{t_1+t_2<t} f(t|t_1,t_2;\lambda,\omega,n_1,n_2)h^{n_1}_{\lambda,\omega}(t_1)h^{n_2}_{\lambda,\omega}(t_2)dt_1dt_2, \quad (5.40)$$

where, as in section 5.1, $h^n_{\lambda,\omega}$ denotes the density of the resultant length. An important special case of (5.40) occurs for $n_2 = 1$, where the above derivations are still valid with the modifications that $T_2 = 1$ and $S_2 = X_{21}$. In this case (5.39) is the conditional density of T given T_1 and (5.40) gives the following recursion formula for $h^n_{\lambda,\omega}$

$$h^n_{\lambda,\omega}(t) = \int_{n-1}^{t-1} f(t|t_1,1;\lambda,\omega,n-1,1)h^{n-1}_{\lambda,\omega}(t_1)dt_1 \ . \quad (5.41)$$

Since we know $h^n_{\lambda,\omega}$ for $n = 2$ (cf. (5.10)) we may calculate $h^n_{\lambda,\omega}$ recursively from (5.41), at least numerically, for very small values of n, perhaps, say, $n = 3$ or 4.

We now consider the case $\lambda = \pm 1/2$, where the distribution of the resultant length is known. From (5.7) and (5.8) we have

$$h^n_{1/2,\omega}(t) = \frac{(t^2-n^2)^{\frac{n-3}{2}} \, n \, K_{n/2}(t\omega)}{t^{n/2-1}\,\Gamma(\frac{n-1}{2})\sqrt{\pi}\; 2^{n/2-2}\omega^{-n/2}e^{-n\omega}} \qquad (t>n).$$

Using (5.39) with $\lambda = 1/2$, we obtain the following density for $T_1,T_2|T$

$$\frac{\left(\frac{u^{1/2}t_1+u^{-1/2}t_2}{u^{-1/2}t_1+u^{1/2}t_2}\right)^{\frac{n.\lambda}{2}} u^{\frac{(n_2-n_1)\lambda}{2}} + \left(\frac{u^{1/2}t_1+u^{-1/2}t_2}{u^{-1/2}t_1+u^{1/2}t_2}\right)^{-\frac{n.\lambda}{2}} u^{-\frac{(n_2-n_1)\lambda}{2}}}{\sqrt{(t^2-t_1^2-t_2^2)^2-4t_1^2t_2^2}}$$

$$\times \frac{(t_1^2-n_1^2)^{\frac{n_1-3}{2}}(t_2^2-n_2^2)^{\frac{n_2-3}{2}} n_1 n_2 t^{n./2}\,\Gamma(\frac{n.-1}{2})\,4}{(t^2-n.^2)^{\frac{n.-3}{2}} n.\, t_1^{n_1/2-1} t_2^{n_2/2-1}\,\Gamma(\frac{n_1-1}{2})\Gamma(\frac{n_2-1}{2})\sqrt{\pi}}$$

$$(t_1>n_1,\ t_2>n_2,\ t>t_1+t_2). \qquad (5.42)$$

To obtain the density of $T_1+T_2|T$ for the test we integrate (5.42) over contours of constant values for T_1+T_2.

In the case $\lambda = 0$ the distribution of the resultant length is also known. From section 5.1 we have

$$h^n_{0,\omega}(t) = \frac{2K_0(\omega t)t}{K_0(\omega)^n \pi} B(t,n) \qquad (t>n),$$

where $B(t,n) = \mathrm{Re}[i\int_0^\infty J_0(ts)K_0(is)^n s\,ds]$. Hence the conditional density of T_1,T_2 given T has the form

$$\frac{4\,t_1\,t_2\,B(t_1,n_1)\,B(t_2,n_2)}{\sqrt{(t^2-t_1^2-t_2^2)^2-4t_1^2t_2^2}\;\pi B(t,n.)} \qquad (t_1>n_1,\ t_2>n_2,\ t>t_1+t_2) \qquad (5.43)$$

and again we may integrate (5.43) to get the density for $T_1+T_2|T$.

Let us turn to the problem of estimation in the model where H_0 is not necessarily true. For $\omega>0$ the likelihood is

$L(\omega, \eta_1, \ldots, \eta_k)$

$$= \frac{\prod_{i=1}^{k} \eta_i^{-n_i \lambda}}{2^{n.} K_\lambda(\omega)^{n.}} \prod_{i=1}^{k} \prod_{j=1}^{n_i} x_{ij}^{\lambda-1} e^{-\frac{\omega}{2} \sum_{i=1}^{k} (\eta_i x_{i\dashv} + \eta_i^{-1} x_{i.})},$$

where $n. = \sum_{i=1}^{k} n_i$. For any given $\omega > 0$ it easily follows that L is maximized for

$$\eta_i = \frac{-\frac{\lambda}{\omega} + \sqrt{(\frac{\lambda}{\omega})^2 + \bar{x}_{i.} \bar{x}_{i\dashv}}}{\bar{x}_{i\dashv}} = \frac{\bar{x}_{i.}}{\frac{\lambda}{\omega} + \sqrt{(\frac{\lambda}{\omega})^2 + \bar{x}_{i.} \bar{x}_{i\dashv}}}.$$

The derivative of $\ln L$ with respect to ω is, using (A.3), (A.11) and (A.17),

$$\frac{\partial}{\partial\omega} \ln L = n.\sqrt{(\frac{\lambda}{\omega})^2 + D_\lambda(\omega)} - \frac{1}{2} \sum_{i=1}^{k} (\eta_i x_{i\dashv} + \eta_i^{-1} x_{i.}),$$

and hence stationary points of L are determined by the equation

$$n.\sqrt{(\frac{\lambda}{\omega})^2 + D_\lambda(\omega)} - \sum_{i=1}^{k} n_i \sqrt{(\frac{\lambda}{\omega})^2 + \bar{x}_{i.} \bar{x}_{i\dashv}} = 0. \tag{5.44}$$

For $\lambda = 0$ (5.44) turns into the equation

$$\sqrt{D_0(\omega)} = \frac{1}{n.} \sum_{i=1}^{k} T_i \tag{5.45}$$

which has a unique solution that corresponds to a maximum for L, except if $T_i = n_i$, $i = 1, \ldots, k$. In the case $\lambda = 0$ we furthermore have the following simple maximum likelihood estimates for η_i,

$$\hat{\eta}_i = S_i, \qquad i = 1, \ldots, k.$$

For $\lambda \neq 0$ we are unable to give a complete description of the stationary points for L, but the following points can be made: An equivalent version of (5.44) is

$$\frac{1}{n.} \sum_{i=1}^{k} n_i \, f_i(\omega) = 1,$$

where

$$f_i(\omega) = \frac{\sqrt{(\frac{\lambda}{\omega})^2 + \bar{x}_{i.}\bar{x}_{i\dashv}}}{\sqrt{(\frac{\lambda}{\omega})^2 + D_\lambda(\omega)}} .$$

From (4.13) and (A.20) we have

$$\lim_{\omega \downarrow 0} f_i(\omega) = 1,$$

and from (A.22)

$$\lim_{\omega \to \infty} f_i(\omega) = \sqrt{\bar{x}_{i.}\bar{x}_{i\dashv}} \geq 1.$$

From Theorem 4.1 we know that if

$$|\lambda| \leq \bar{x}_{i.}\bar{x}_{i\dashv}/(\bar{x}_{i.}\bar{x}_{i\dashv} - 1) \tag{5.46}$$

then the equation $f_i(\omega) = 1$ has a solution, in which case it follows that f_i is non-monotone. In the case where the converse of (5.46) holds for all i (this can only be the case for $|\lambda| > 1$) the left hand side of (5.44) is negative for any $\omega > 0$, and hence $\hat{\omega} = 0$.

An ad hoc estimate for ω with a certain intuitive appeal is obtained as the solution to the equation

$$\sqrt{D_\lambda(\omega)} = \frac{1}{n.} \sum_{i=1}^{k} T_i,$$

the analogue of (5.45). For small values of $|\lambda|$ and large values of ω this will be close to a solution of (5.44), and may eventually serve as a starting value if (5.44) is to be solved iteratively.

5.5. A regression model

Assume that $X_1,\dots,X_n$ are independent random variables with distributions

$$X_i \sim N^{-}(\lambda,\omega\eta_i,\omega\eta_i^{-1}), \qquad i = 1,\dots,n .$$

A simple regression model is obtained by letting η_i be proportional to some known function $h>0$:

$$\eta_i = \eta_0 h(t_i),$$

where $\eta_0>0$ is an unknown parameter and $t_1,\dots,t_n$ are known covariates.

For given ω and λ the model in fact specifies the mean value as proportional to h

$$EX_i = R_\lambda(\omega)\eta_0 h(t_i).$$

From (2.5) it follows that the variables $X_i h(t_i)^{-1}$ are i.i.d. with distribution $N^{-}(\lambda,\omega\eta_0,\omega\eta_0^{-1})$, and hence we may estimate ω and η_0 in the usual way (cf. Theorem 4.1).

One may extend this model by letting h depend on some unknown parameter θ, but there appears to be no non-trivial choice for h which makes the model mathematically convenient.

6. The hazard function. Lifetime models.

In this chapter we shall first give some comments on the interpretation of the generalized inverse Gaussian distribution as a lifetime model, and in section 6.1 we shall investigate the shape of its hazard function.

It was shown by Barndorff-Nielsen, Blæsild and Halgreen (1978) that the generalized inverse Gaussian distribution with $\lambda \leq 0$ is a first hitting time distribution for certain time-homogeneous diffusion processes, suggesting its potential use as a lifetime distribution or the distribution for times between events in a renewal process.

In this respect the inverse Gaussian is also central, being the first hitting time distribution for a Brownian motion with drift. In fact it originally appeared in this context, see Schrödinger (1915). If the Brownian motion is drift-free the associated first hitting time distribution is $N^{-}(-\frac{1}{2}, \chi, 0)$, i.e. the one-sided stable law with exponent 1/2.

This interpretation of the inverse Gaussian as a first hitting time distribution has played a central role in several applications where one was actually observing waiting times (see Chhikara and Folks, 1978). For example, Lancaster (1972) used the inverse Gaussian distribution as a model for the duration of strikes.

In demography the inverse Gaussian distribution is known as the "Hadwiger curve" (Hadwiger (1940) , Keyfitz (1968, p. 149-168)) and is used, in a suitable form, as a "net maternity function", i.e. a curve describing the fertility of a mother as a function of time. Hadwiger (1940) studied the renewal process which is generated by specifying the net maternity function to be a constant times the density function of the inverse Gaussian distribution, and he derived a convolution formula which is in effect (2.12). However, until quite recently (Hoem, 1976 p.182) there appears to have been no interaction between the development of the inverse Gaussian distribution in the two fields (statistics and demography); in particular there seems to be no connection between the works of Tweedie (1957) and Hadwiger (1940).

Unfortunately, the generalized inverse Gaussian distribution appears not to be suitable for use in realiability and lifetesting situations with censored data, because the survivor function is not available in simple form, not even for the inverse Gaussian distribution where it can be expressed only in terms of the normal distribution function (cf. Chhikara and Folks, 1978). For the numerical work presented below we have used numerical integration of the density, but it could be noted that Faxén (1920) has considered a series expansion of an integral of the kind needed here.

In order to gain some insight into the type of failure mechanisms that we might think of as giving rise to lifetime distributions that are generalized inverse Gaussian we shall now examine the hazard function of the distribution.

6.1. Description of the hazard function

Let f be the probability density function (1.1) of the generalized inverse Gaussian distribution and let $\bar{F}(x) = \int_x^\infty f(t)dt$ and $r(x) = f(x)/\bar{F}(x)$ be the corresponding survivor function and hazard function, respectively.

The description of the hazard function for the generalized inverse Gaussian distribution is as follows (the proof is given at the end of this section): For $\lambda < 1$ the hazard is unimodal with zero initial value and asymptotic value $\psi/2$. The mode point m_r of the hazard satisfies the inequalities

$$\frac{\chi}{2(1-\lambda)} \leq m_r \leq \frac{\chi}{1-\lambda} . \tag{6.1}$$

The gamma distribution $(\chi = 0)$ with $0 < \lambda < 1$ is a degenerate case where the hazard is infinite at the origin and decreases towards the asymptotic value $\psi/2$. For the reciprocal gamma distribution $(\psi = 0,\ \lambda < 0)$ we see that the hazard has zero asymptotic value. In the case $\lambda \geq 1$ the hazard has zero initial value and increases towards the asymptotic value $\psi/2$, except for the exponential distribution $(\chi = 0,\ \lambda = 1)$ whose hazard is constant and equal to $\psi/2$. Figure 6.1 shows plots of the hazard function for various values of the parameters.

The unimodal behaviour of the hazard in the case $\lambda < 1$, $\chi > 0$ indicates an initial burn-in period in the underlying failure mechanism, i.e. a high occurrence of early failures. If we look at the form of the density in this case we see that the factor $x^{\lambda-1}$ gives a peak near zero which is damped by the factor $e^{-\frac{1}{2}\chi x^{-1}}$. For $\psi > 0$ the factor $e^{-\frac{1}{2}\psi x}$ dominates the tail of the density, giving rise to the asymptotic constant value of the hazard. For $\psi = 0$, $\lambda < 0$ the density has a Pareto-like (geometric) tail $x^{\lambda-1}$, but near zero the density does not resemble the Pareto density in this case, except for small values of χ.

The hazard function of the log normal distribution also has a unimodal form, but has always zero asymptotic value. The Weibull distribution and the generalized Pareto distribution (Davis and Feldstein, 1979) both have monotone hazards and have survivor functions that are available in useful form, allowing simple analysis of censored data, an advantage which, as noted earlier, is not shared by the generalized inverse Gaussian distribution.

Since $r(x) = -\frac{\partial}{\partial x} \ln \bar{F}(x)$ we have that the log-survivor function $\ln \bar{F}$ is concave (convex) when r is increasing (decreasing), and hence we have a good idea of the shape of the log-survivor function. Let us just note that for $\lambda < 1$, $\chi > 0$, $\ln \bar{F}$ is concave for $x < m_r$ and convex for $x > m_r$, where m_r is the mode point of the hazard function. For $\psi > 0$ the tail of $\ln \bar{F}$ is linear with slope $-\psi/2$, whereas for $\psi = 0$, $\lambda < 0$ the tail has the form $\lambda \ln x$ (the same form as for the log normal distribution). Figure 6.2 shows plots of the log-survivor function for some values of the parameters. For all plots we have chosen the same value for ψ in order to make the slope of the tail of $\ln \bar{F}$ the same (the case $\psi = 0$ is shown in a separate plot).

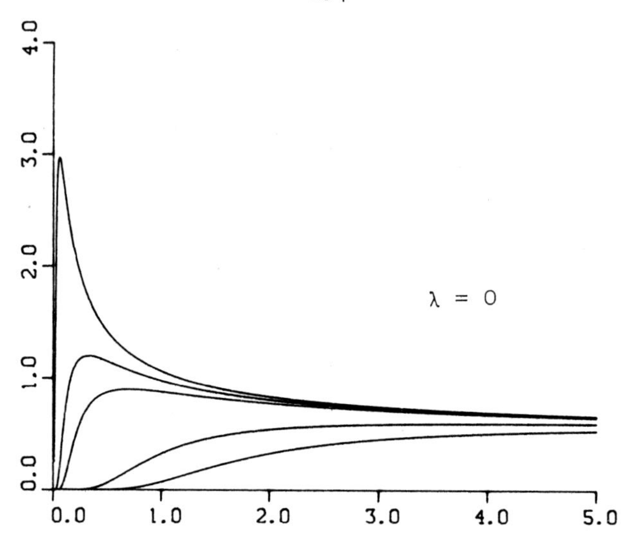

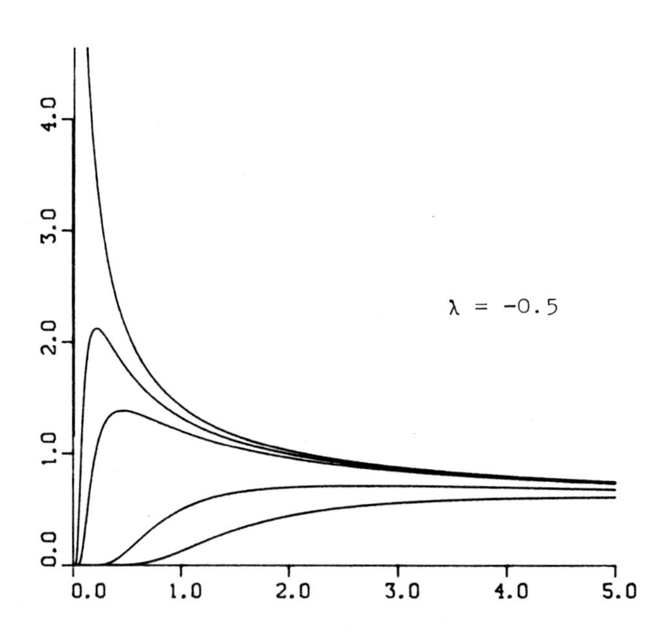

Figure 6.1 A. Plots of hazard function for $\psi = 1$, $\chi = 0.1$, 0.5, 1, 5, 10. The hazard decreases as a function of χ.

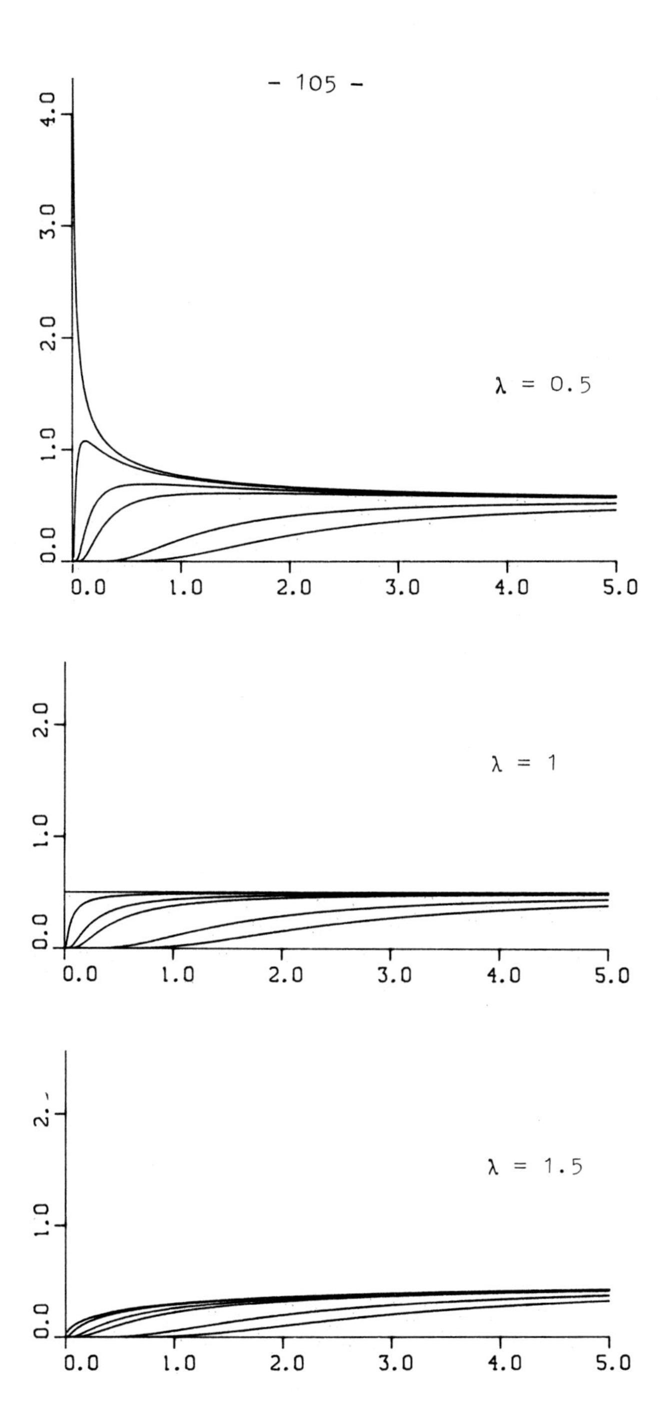

Figure 6.1 B. Plots of hazard function for $\psi = 1$, $\chi = 0, 0.1, 0.5, 1, 5, 10$. The hazard decreases as a function of χ.

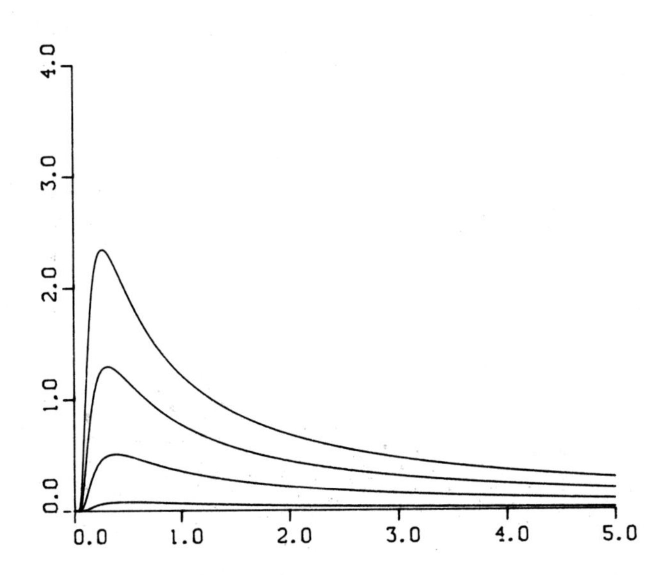

Figure 6.1 C. Plots of hazard function for $\psi = 0$, $\chi = 1$ and $\lambda = -0.1, -0.5, -1, -1.5$. The hazard decreases as a function of λ.

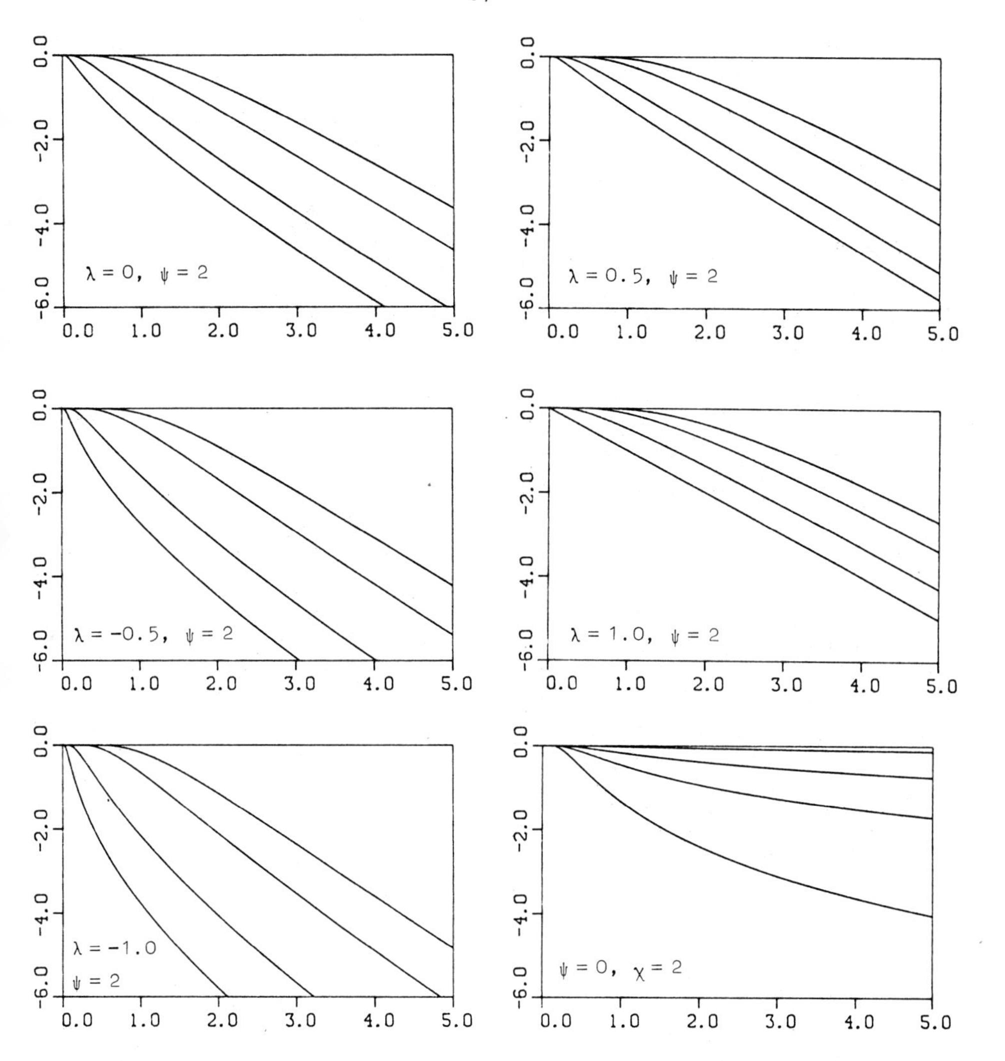

Figure 6.2. Plots of log-survivor function. For $\psi = 2$, χ takes the values 0, 1, 4, 8 (if $\lambda > 0$) or 0.25, 1, 4, 8 (if $\lambda \leq 0$). (The log-survivor function increases with χ.) For $\psi = 0$, λ takes the values -0.1, -0.5, -1.0 and -2.0 (the log-survivor function increases with λ).

Proof of assertions about the hazard function

We shall now give the proof of our previous assertions about the hazard function (at the beginning of this section). Our method of proof is a simple extension of the technique used by Chhikara and Folks (1977), who examined the hazard function of the inverse Gaussian distribution.

The proof falls in two parts; the first concerns the monotonicity properties of the hazard function and the second the asymptotic value of the hazard.

For the proof of the monotonicity properties we need to consider the cases $\lambda < 1$ and $\lambda \geq 1$ separately, but we first define some functions that we need in both cases.

Let the functions p and g be defined by

$$p(x) = -\frac{\partial}{\partial x}\ln f(x) = -(\lambda-1)x^{-1} - \frac{1}{2}\chi x^{-2} + \frac{1}{2}\psi$$

and

$$g(x) = \int_x^\infty \frac{f'(t)}{p(t)}dt + \frac{f(x)}{p(x)} = -\bar{F}(x) + \frac{f(x)}{p(x)},$$

respectively. Let m denote the mode of f, i.e. the zero of p (cf. chapter 2):

$$m = \begin{cases} \dfrac{\lambda-1+\sqrt{(\lambda-1)^2+\chi\psi}}{\psi} & \text{if } \psi > 0 \\[2ex] \dfrac{\chi}{2(1-\lambda)} & \text{if } \psi = 0 \end{cases}$$

and note that $p(x) < 0$ (> 0) for $x < m$ $(x > m)$. The derivative of p is

$$p'(x) = (\lambda-1)x^{-2} + \chi x^{-3}$$

and the logarithmic derivative of r is

$$\frac{r'(x)}{r(x)} = \frac{p(x)}{\bar{F}(x)} g(x). \tag{6.2}$$

Figure 6.3 illustrates the functions f, r and p in the case $\lambda < 1$, $\chi > 0$.

$\underline{\lambda < 1}$

In this case p has mode $m_p = \chi/(1-\lambda)$ and $m \leq m_p$. Now, for $x < m$ f is increasing and $\bar{F}$ is decreasing and hence r is increasing. For $x > m_p$ p is decreasing and hence

$$g(x) < \int_x^\infty \frac{f'(t)}{p(x)} dt + \frac{f(x)}{p(x)} = 0.$$

Noting that $p(x) > 0$ for $x > m$ it follows from (6.2) that r is decreasing for $x > m_p$.

Thus we have that $r'(x)/r(x) > 0$ for $x < m$ and $r'(x)/r(x) < 0$ for $x > m_p$ and hence r'/r, being continuous, has at least one zero between m and m_p. Now

$$g'(x) = -\frac{f(x)p'(x)}{p(x)^2} < 0 \quad \text{for} \quad x < m_p,$$

and hence it follows from (6.2) that $r'(x)/r(x)$ can have at most one zero for $m < x < m_p$, since both $\bar{F}$ and p are positive in this interval.

We conclude that r has a unique mode m_r satisfying $m \leq m_r \leq m_p$. Since $\frac{\partial}{\partial x} \ln r(x) = -p(x) + r(x)$ m_r is the solution to the equation

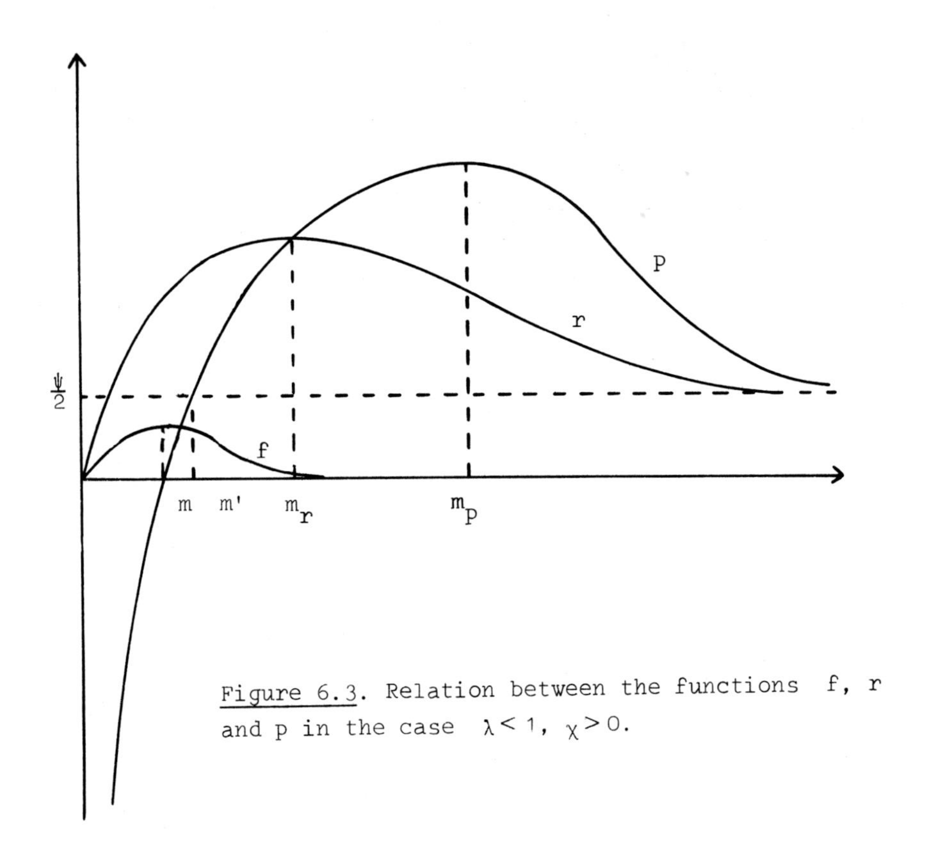

Figure 6.3. Relation between the functions f, r and p in the case $\lambda < 1$, $\chi > 0$.

$$p(m_r) = r(m_r). \tag{6.3}$$

Now, letting $m' = \chi/2(1-\lambda)$ one has $m \leq m' \leq m_p$ and $p(m') = \psi/2$. Since $\lim_{x\to\infty} p(x) = \psi/2$ and (as will be shown later) $\lim_{x\to\infty} r(x) = \psi/2$ we conclude from (6.3) that $m' \leq m_r$, whence (6.1) has been proved. The inequality $m_r \geq m'$ was in the case $\lambda = -\frac{1}{2}$ given in the discussion of Chhikara and Folks (1978) by Professor A.Lancaster.

$\lambda \geq 1$

For the exponential distribution $(\lambda = 1, \chi = 0)$ we have $r \equiv \psi/2$. Otherwise we can argue as follows: For $x < m$ we have, by the same argument as for $\lambda < 1$, that r is increasing. Now, since $p'(x) > 0$ for any $x > 0$ we have

$$g'(x) = -\frac{f(x)p'(x)}{p(x)^2} < 0 \quad (x \neq m).$$

Noting that $\lim_{x\to\infty} p(x) = \psi/2$ we have from the definition of g that $\lim_{x\to\infty} g(x) = 0$. Thus $g(x) > 0$ for $x > m$, and using (6.2) it follows that $r'(x)/r(x) > 0$ for $x > m$. We conclude that r is increasing for $x > 0$.

The behaviour of the hazard at the origin is examined by noting that for $\chi = 0$, $0 < \lambda < 1$ we have $f(0^+) = \infty$ and for $\chi = 0$, $\lambda = 1$ we have $f(0^+) = \frac{\psi}{2}$ and in all other cases $f(0^+) = 0$. This completes the first part of the proof.

For the second part of the proof we derive some inequalities for $r(x)$ that allow us to conclude

$$\lim_{x\to\infty} r(x) = \psi/2. \tag{6.4}$$

First, note that

$$\frac{1}{r(x)} = e^{\frac{\chi}{2}x^{-1} + \frac{\psi}{2}x} \int_x^\infty \left(\frac{t}{x}\right)^{\lambda-1} e^{-\frac{1}{2}(\chi t^{-1} + \psi t)} dt \tag{6.5}$$

$$= \int_x^\infty \left(\frac{t}{x}\right)^{\lambda-1} e^{-\frac{1}{2}(\chi(t^{-1} - x^{-1}) + \psi(t-x))} dt.$$

The last expression shows that r is monotone as a function of each of the parameters λ, χ and ψ for fixed value of x and the remaining two parameters, in fact increasing as a function of ψ and decreasing as a function of λ and χ. In general we have that for an exponential family of the form $a(\theta)b(x)e^{\theta t(x)}$ the hazard is a monotone function of θ for fixed x if $t(x)$ is a monotone function of x.

Substituting $t = xy$ in (6.5) we get

$$\frac{1}{r(x)} = e^{\frac{\chi}{2}x^{-1} + \frac{\psi}{2}x} \int_1^\infty y^{\lambda-1} e^{-\frac{1}{2}(\frac{\chi}{xy} + \psi x y)} x\, dy.$$

For any $x > 0$, $y > 1$ we have

$$e^{-\frac{\chi}{2x}} \le e^{-\frac{\chi}{2xy}} \le 1,$$

and hence

$$e^{\frac{\psi}{2}x} \int_1^\infty y^{\lambda-1} e^{-\frac{1}{2}\psi xy} x\, dy \le \frac{1}{r(x)} \le e^{\frac{\chi}{2}x^{-1} + \frac{\psi}{2}x} \int_1^\infty y^{\lambda-1} e^{-\frac{1}{2}\psi xy} x\, dy. \quad (6.6)$$

Now we have to distinguish between the case where $\psi > 0$ and the case where $\psi = 0$.

$\underline{\psi > 0}$

In this case we have, integrating by parts,

$$\int_1^\infty y^{\lambda-1} e^{-\frac{1}{2}\psi xy} x\, dy = \frac{2}{\psi}(e^{-\frac{1}{2}\psi x} + (\lambda-1) \int_1^\infty y^{\lambda-2} e^{-\frac{1}{2}xy} dy).$$

Thus, letting

$$A(x) = 1 + (\lambda-1) \int_1^\infty y^{\lambda-2} e^{-\frac{1}{2}\psi x(y-1)} dy$$

(the corresponding formula in Chhikara and Folks (1977) is in error)

(6.6) turns into

$$\frac{\psi}{2}A(x)^{-1}e^{-\frac{\chi}{2}x^{-1}} \leq r(x) \leq \frac{\psi}{2}A(x)^{-1}.$$

Clearly $\lim_{x\to\infty} A(x) = 1$, whence (6.4) follows. One notes that for $\lambda < 1$

$$0 \leq (1-\lambda)\int_1^\infty y^{\lambda-2}e^{-\frac{1}{2}\psi x(y-1)}dy \leq (1-\lambda)\int_1^\infty y^{\lambda-2}dy = 1,$$

and hence $0 \leq A(x) \leq 1$, whereas for $\lambda \geq 1$ we have $A(x) \geq 1$.

$\psi = 0 \quad (\lambda < 0)$

In this case (6.6) turns into

$$-\frac{\lambda}{x}e^{-\frac{\chi}{2}x^{-1}} \leq r(x) \leq -\frac{\lambda}{x},$$

whence (6.4) follows. This completes the proof.

7. Examples

In this chapter we illustrate some of the methods developed in the preceding chapters by analyzing a few sets of data, using the generalized inverse Gaussian distribution.

The first three sets of data are taken from the monograph by Cox and Lewis (1966) on the statistical analysis of series of events, and the original references may be found there. These data can all be considered as realizations of some kind of point process on the real line, the intervals between successive events possibly being dependent. However, we assume for all three examples that the data come from a generalized inverse Gaussian renewal process, the analysis by Cox and Lewis having shown no clear evidence against the renewal hypothesis. Thus we implicitly assume stationarity of the series, and we have not considered the question of testing for trend.

For the nerve pulse data (section 7.2) and the airconditioning equipment data (section 7.1) it is of interest to examine whether there are departures from the exponential distribution, since Cox and Lewis mainly used this distribution in their analysis of these data, whereas for the traffic data (section 7.3) it is of interest to see whether the generalized inverse Gaussian renewal model can explain the data in as good a way as the more complicated models with dependence between intervals that Cox and Lewis employed.

The data that we consider in section 7.4 and 7.5 have been analyzed by Chhikara and Folks (1977,1978) using the inverse Gaussian distribution (the original references may be found in those papers). For the first set of data the distribution is not far from being inverse Gaussian, whereas for the last one has an example where the data is very uninformative about the values of the parameters in the generalized inverse Gaussian distribution.

7.1 Failures of airconditioning equipment

The data in Table 7.1 are the intervals in operating hours between successive failures of airconditioning equipment in 13 Boeing 720 aircraft. Aircraft 11 with only two observations is, however, omitted from the following analysis.

Our basic model is that the intervals between successive failures are independent and have a generalized inverse Gaussian distribution and that, for aircraft no. i, the intervals have the distribution $N^{-}(\lambda^{(i)},\chi^{(i)},\psi^{(i)})$. Table 7.2 shows the maximum likelihood estimates for parameters in this model and estimates of their standard deviations. The ideal would have been to invert

Aircraft												
1	2	3	4	5	6	7	8	9	10	11	12	13
194	413	90	74	55	23	97	50	359	50	130	487	102
15	14	10	57	320	261	51	44	9	254	493	18	209
41	58	60	48	56	87	11	102	12	5		100	14
29	37	186	29	104	7	4	72	270	283		7	57
33	100	61	502	220	120	141	22	603	35		98	54
181	65	49	12	239	14	18	39	3	12		5	32
	9	14	70	47	62	142	3	104			85	67
	169	24	21	246	47	68	15	2			91	59
	447	56	29	176	225	77	197	438			43	134
	184	20	386	182	71	80	188				230	152
	36	79	59	33	246	1	79				3	27
	201	84	27	15	21	16	88				130	14
	118	44	153	104	42	106	46					230
	34	59	26	35	20	206	5					66
	31	29	326		5	82	5					61
	18	118			12	54	36					34
	18	25			120	31	22					
	67	156			11	216	139					
	57	310			3	46	210					
	62	76			14	111	97					
	7	26			71	39	30					
	22	44			11	63	23					
	34	23			14	18	13					
		62			11	191	14					
		130			16	18						
		208			90	163						
		70			1	24						
		101			16							
		208			52							
					95							

Table 7.1. Numbers of operating hours between successive failures of airconditioning equipment in 13 aircraft.

Aircr. i	n_i	$\hat{\lambda}^{(i)}$	$\hat{\chi}^{(i)}$	$V(\hat{\chi}^{(i)})^{\frac{1}{2}}$	$\hat{\psi}^{(i)}.10^3$	$V(\hat{\psi}^{(i)})^{\frac{1}{2}}.10^3$	$u^{(i)}$	$\hat{\eta}^{(i)}$	$\hat{\omega}^{(i)}$	$\hat{\hat{\omega}}^{(i)}$	$V(\hat{\hat{\omega}}^{(i)})^{\frac{1}{2}}$
1	6	-0.81	77.4	38.6	6.45	9.5	1.78	109.5	0.71	0.52	0.57
2	23	-0.32	38.8	12.8	6.47	3.6	1.47	77.4	0.50	0.47	0.18
3	29	0.37	48.9	20.2	22.66	6.3	2.03	46.4	1.05	1.00	0.31
4	15	-0.90	81.3	23.7	1.81	2.7	1.50	211.8	0.38	0.34	0.31
5	14	1.00	29.0	32.0	18.87	5.7	1.88	39.2	0.74	0.66	0.46
6	30	0.42	3.2	2.2	18.80	5.2	1.23	13.1	0.25	0.23	0.10
7	27	1.13	0.035	0.67	29.34	5.3	1.23	1.1	0.032	0.026	0.32
8	24	0.65	5.4	4.7	24.76	6.5	1.40	14.7	0.36	0.34	0.18
9	9	0.16	1.9	2.3	2.66	1.8	1.04	26.8	0.071	0.056	0.052
10	6	0.07	9.5	9.3	6.31	5.3	1.20	38.7	0.24	0.18	0.16
12	12	0.37	3.5	4.1	9.03	4.1	1.16	19.8	0.18	0.15	0.12
13	16	0.47	48.4	28.1	25.11	9.0	2.10	43.9	1.10	1.01	0.44

Table 7.2. Failures of airconditioning equipment. Maximum likelihood estimates and their standard deviations, basic model (see text). Also shown are the values of $u = \bar{x}.\bar{x}_{\dashv}/(\bar{x}.\bar{x}_{\dashv} - 1)$.

the observed information matrix (4.47) to get an estimate of the asymptotic variance of $(\hat{\lambda}, \hat{\chi}, \hat{\psi})$, but a procedure for calculating (4.47) has not yet been included in our computer program. Hence we have instead inverted the information matrix (4.24), taking $\lambda^{(i)} = \hat{\lambda}^{(i)}$, to get an estimate of the asymptotic variance of $(\hat{\chi}, \hat{\psi})$. The table also shows the maximum likelihood estimate for ω in the marginal distribution of the resultant length, denoted by $\hat{\hat{\omega}}^{(i)}$, and an estimate of the asymptotic standard deviation of $\hat{\hat{\omega}}^{(i)}$, $(ni_{\hat{\lambda}^{(i)}}(\hat{\hat{\omega}}^{(i)}))^{-1/2}$, according to (5.32).

Figure 7.1 shows the (normed) partially maximized log-likelihood $\tilde{l}$ for each of the 12 aircraft. The horizontal line in the plots gives an approximate 95% confidence interval for λ (determined by the intersection points between $\tilde{l}$ and the line), based on the asymptotic $\chi^2(1)$-distribution for $-2\ln Q$ in a test for the hypothesis that λ has a given value. However, the number of observations for each aircraft is too small to assure that the distribution of $-2\ln Q$ actually has its limiting form, particularly since we have to estimate two additional incidental parameters $\chi^{(i)}$ and $\psi^{(i)}$ for each aircraft, but the confidence intervals may give a rough indication of the most reasonable values for λ.

Note here the close connexion between the value of $u^{(i)}$ (Table 7.2) and the shape of $\tilde{l}$. In fact it appears that a value of $u^{(i)}$ which is close to 1 corresponds to a nice parabolic form for $\tilde{l}$ and a narrow confidence interval for $\lambda^{(i)}$, whereas a larger value of $u^{(i)}$ corresponds to an irregular form of $\tilde{l}$

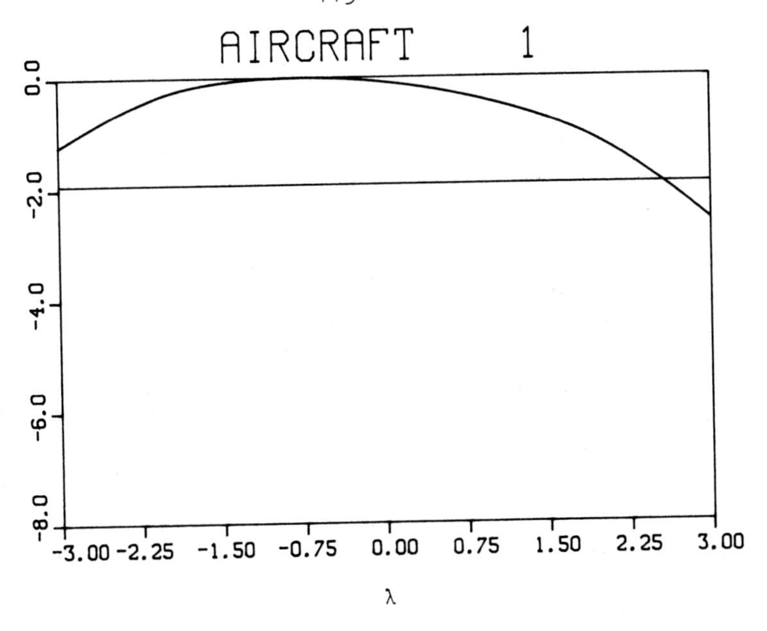

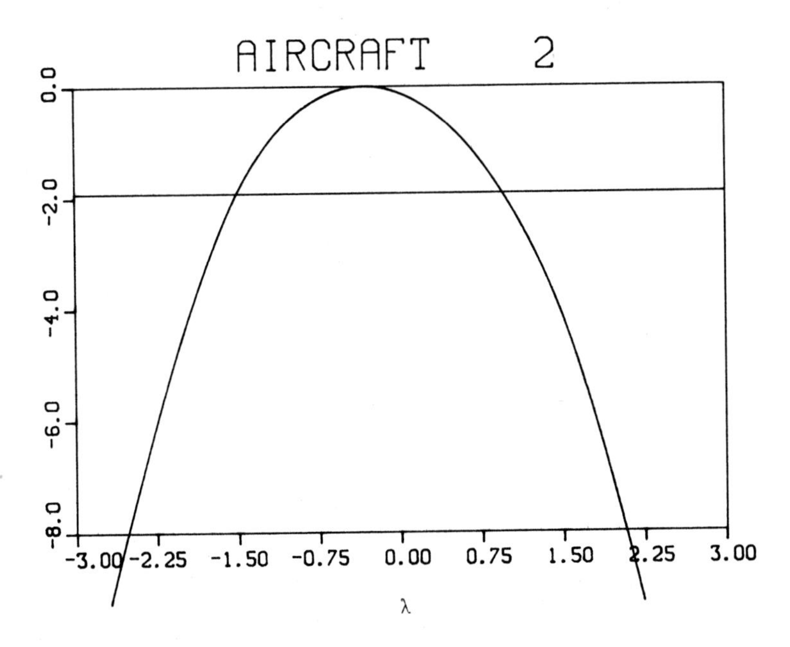

Figure 7.1. (Continued overleaf.) Failures of airconditioning equipment. Plots of the (normed) partially maximized log-likelihood for each of the 12 aircraft.

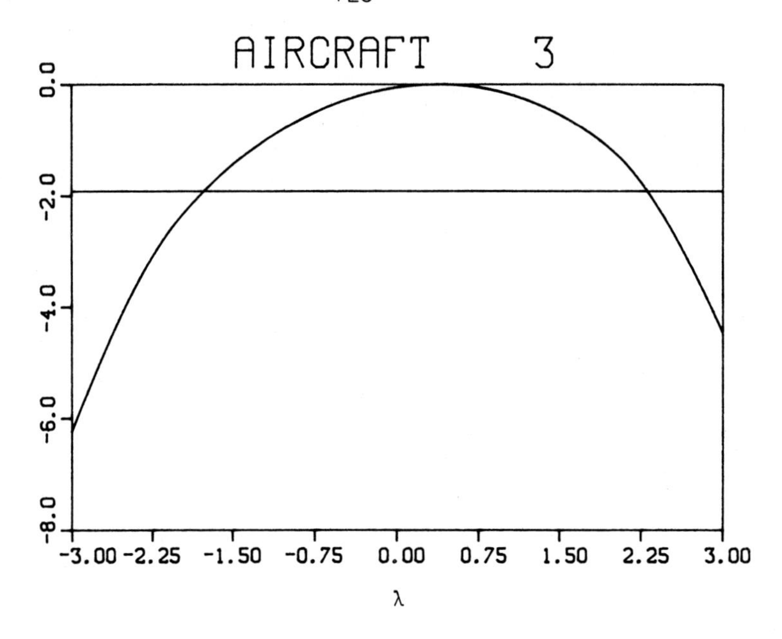

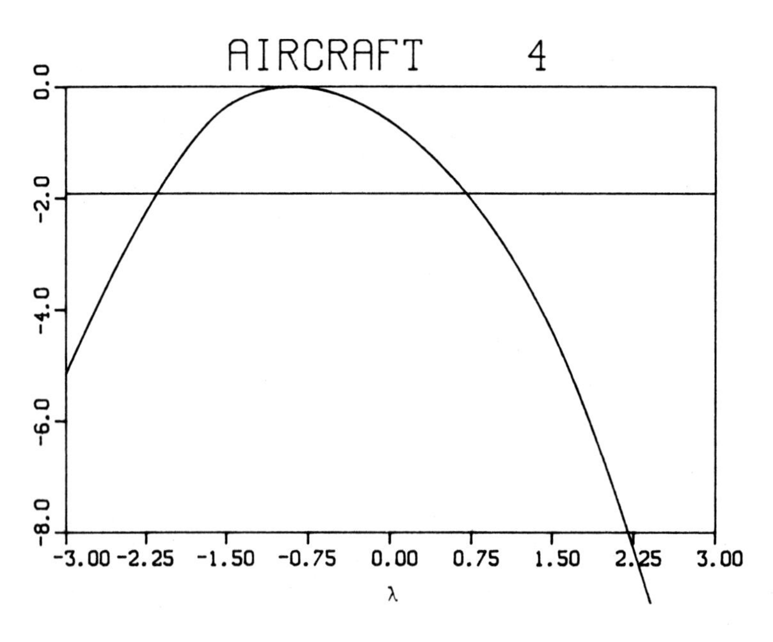

Figure 7.1. (Continued.)

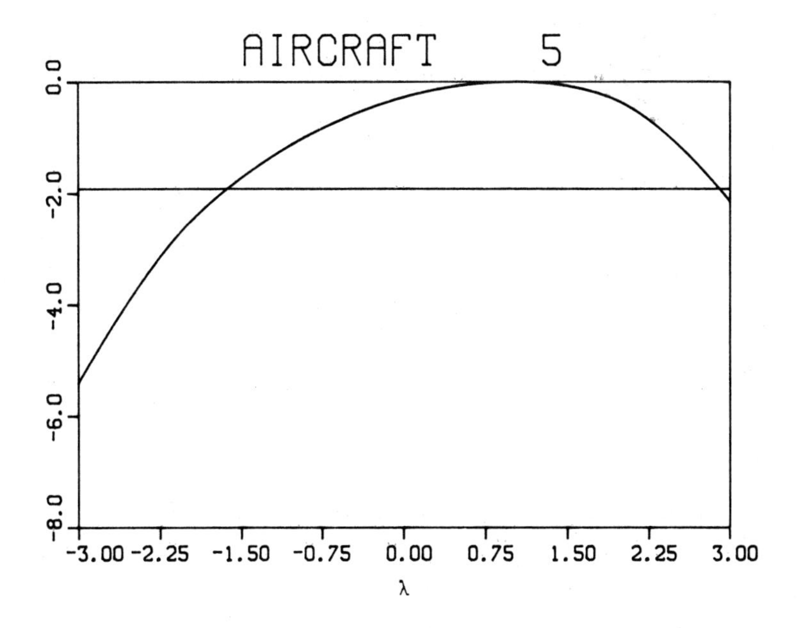

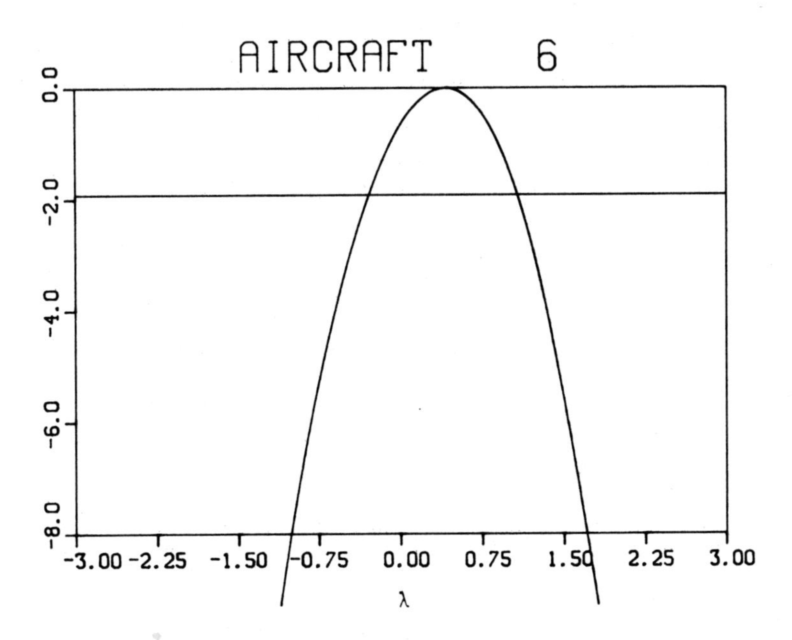

Figure 7.1. (Continued.)

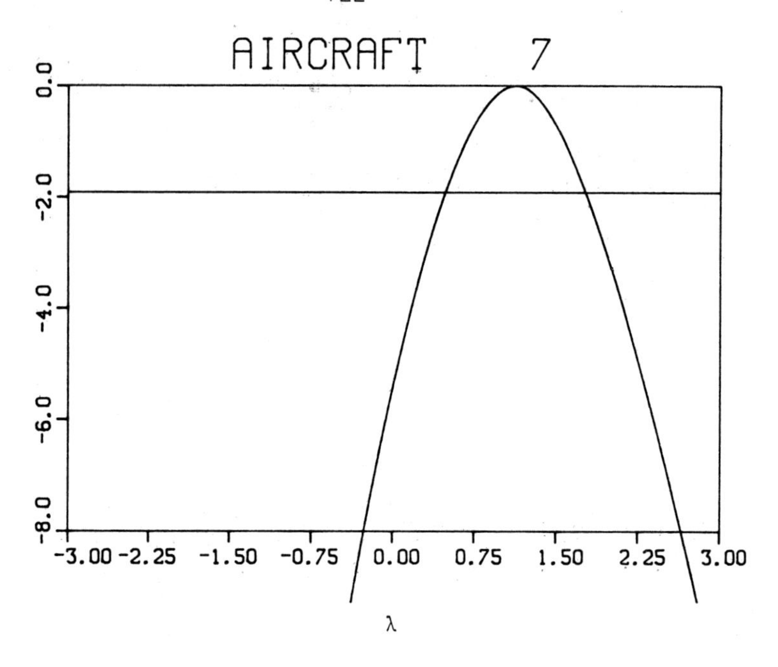

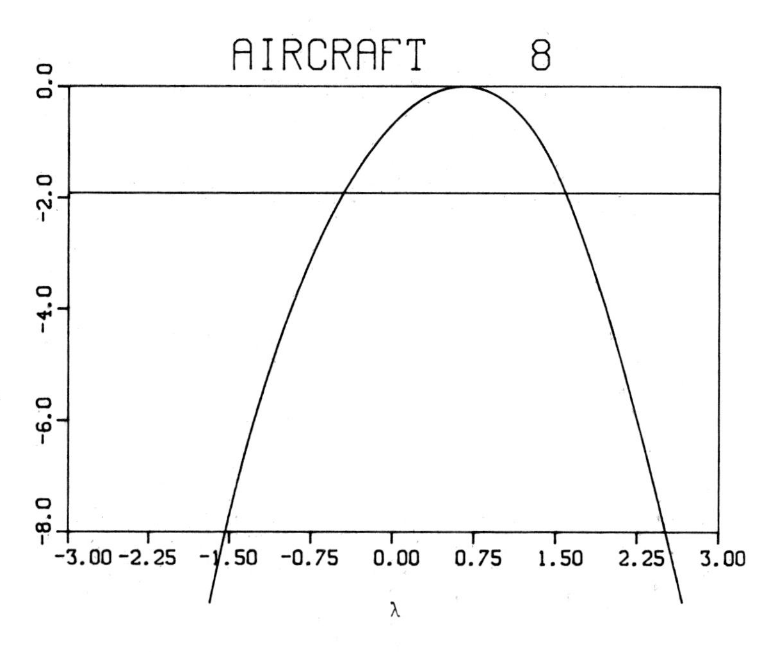

Figure 7.1. (Continued.)

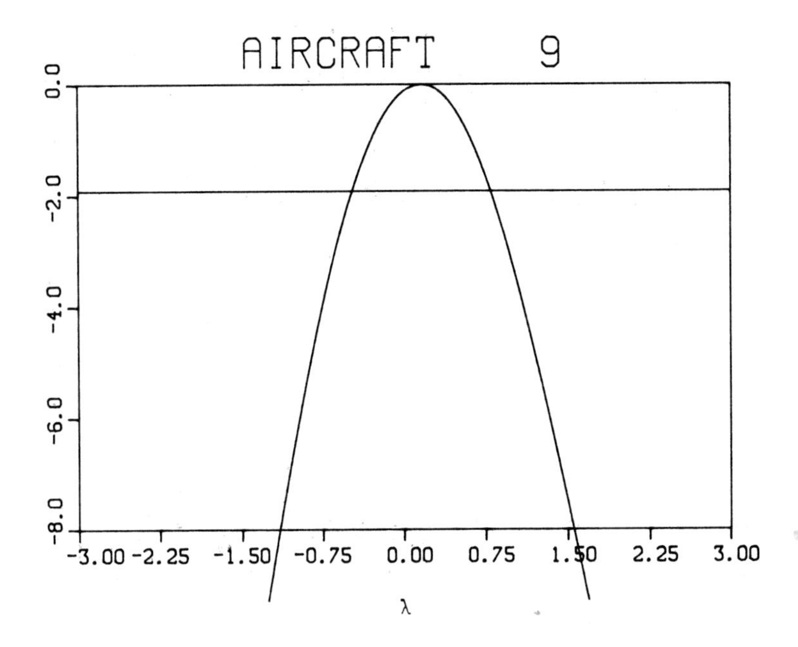

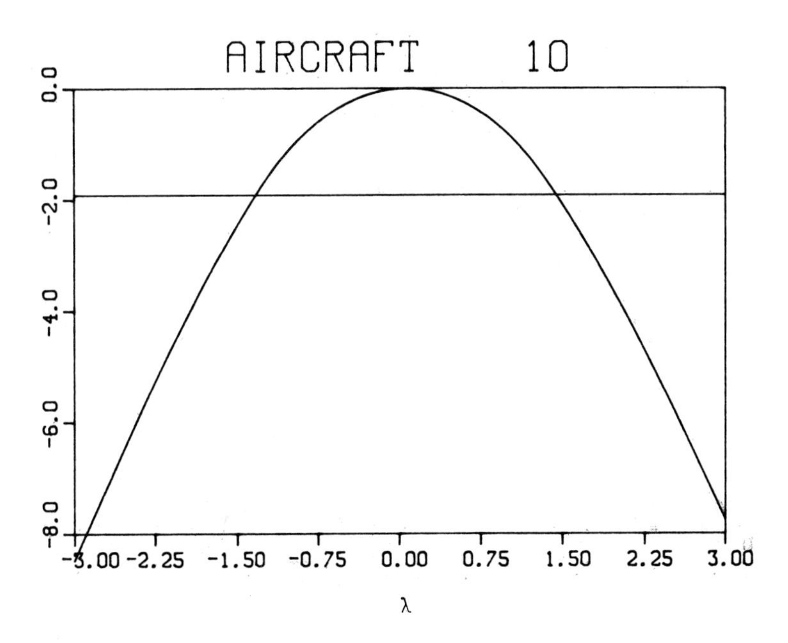

Figure 7.1. (Continued.)

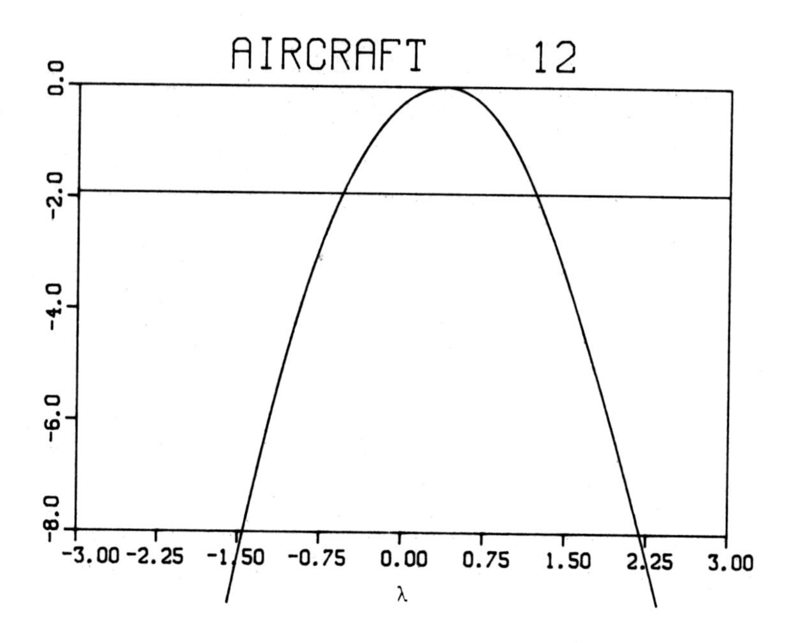

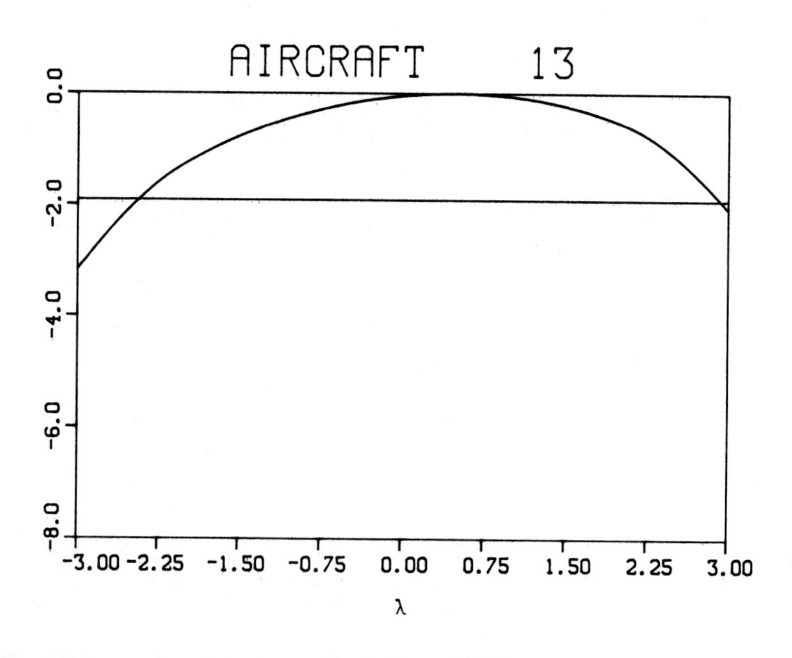

Figure 7.1. (Continued.)

(compare with Figure 4.3) and a wide confidence interval for λ. Cf. the discussion of $\tilde{1}$ in section 4.3.

There are now two main questions to be asked: 1) are there departures from the exponential distribution (or from the gamma distribution) for any of the 12 aircraft, and 2) is it reasonable to assume a common value for λ, and if so, which value could the common λ have. In turn we shall try to answer these questions.

First we consider the question of departures from the exponential distribution for each of the 12 aircraft separately. (Recall that the exponential distribution corresponds to the parameter values $\lambda = 1$, $\chi = 0$). We have to proceed stepwise, first asking whether $\lambda^{(i)} = 1$ and then asking whether $\chi^{(i)} = 0$ (or, equivalently, $\omega^{(i)} = 0$).

The confidence intervals for $\lambda^{(i)}$ in Figure 7.1 do not all contain the value $\lambda = 1$, and $\hat{\lambda}^{(i)}$ is quite systematically smaller than 1. But let us try to test $\omega^{(i)} = 0$, assuming that $\lambda^{(i)} = 1$.

In section 5.3 we argued that this hypothesis should be tested in the marginal distribution of the resultant length. Table 7.3 shows the maximum likelihood estimates $\hat{\hat{\omega}}_1^{(i)}$ in this distribution and the estimated variance in the asymptotic normal distribution of $\hat{\hat{\omega}}_1^{(i)}$ (cf. (5.32)). If we test at level 5% in the asymptotic normal distribution of $\hat{\hat{\omega}}_1^{(i)}$ (large values being significant) we find departures only for aircraft 3 and 13. For these two aircraft the same test applied for $\lambda^{(i)} = \hat{\lambda}^{(i)}$ instead of for $\lambda^{(i)} = 1$ (see Table 7.2) also shows a significant departure from the hypothesis $\omega^{(i)} = 0$, so the two aircraft seem to show a departure from the gamma distribution. Note that the estimates of $\lambda^{(i)}$ for the two aircraft are 0.37 and 0.47, respectively, being close to $\lambda = 0.5$.

For the other 10 aircraft there seem to be no gross departures from exponentiality, although, as mentioned earlier, $\hat{\lambda}$ is quite systematically smaller than 1.

Aircraft i	$\hat{\omega}_1^{(i)}$	$(ni_1(\hat{\omega}_1^{(i)}))^{-1}$	$\sqrt{ni_1(\hat{\omega}_1^{(i)})}\hat{\omega}_1^{(i)}$
1	0.47	0.42	0.73
2	0.29	0.073	1.07
3	0.86	0.13	2.48
4	0.30	0.12	0.87
5	0.66	0.21	1.44
6	0.075	0.039	0.38
7	0.076	0.044	0.36
8	0.22	0.063	0.88
9	$1.7\cdot 10^{-6}$	0.14	$4.54\cdot 10^{-6}$
10	0.031	0.25	0.062
12	0.021	0.10	0.066
13	0.88	0.24	1.80

Table 7.3. Maximum likelihood estimates $\hat{\omega}_1^{(i)}$ in the marginal distribution of the resultant length (for $\lambda = 1$) and estimates of their variances.

It seems, however, more efficient to proceed by first examining whether there is a common value for λ and then to examine whether this common value can be 1.

The confidence intervals for $\lambda^{(i)}$ in Figure 7.1 do in fact indicate that there is a common value (the intersection between the 12 intervals is the interval $0.5 < \lambda < 0.75$) and the likelihood ratio test ($-2\ln Q = 10.16$, 11 degrees of freedom) also supports the hypothesis.

A graphical check of the hypothesis is obtained by plotting $\hat{\omega}^{(i)}$ or $\hat{\eta}^{(i)}$ against $\hat{\lambda}^{(i)}$ and then looking for any unexpected systematic effects. Such plots are shown in Figure 7.2 and 7.3.

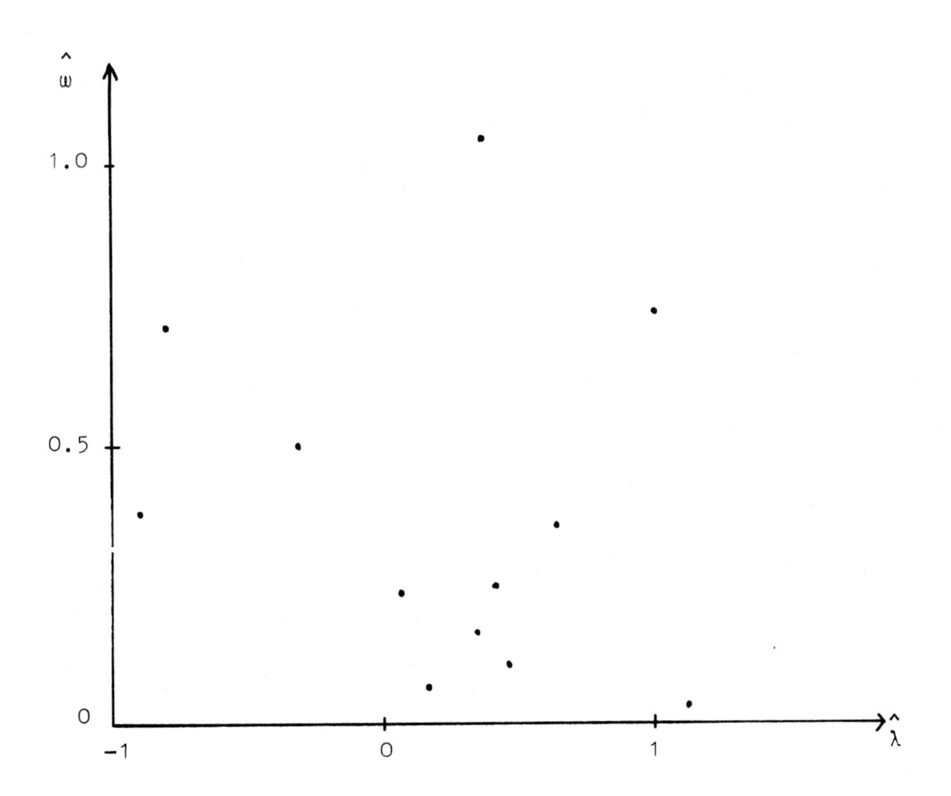

Figure 7.2. Plot of $\hat{\omega}^{(i)}$ against $\hat{\lambda}^{(i)}$.

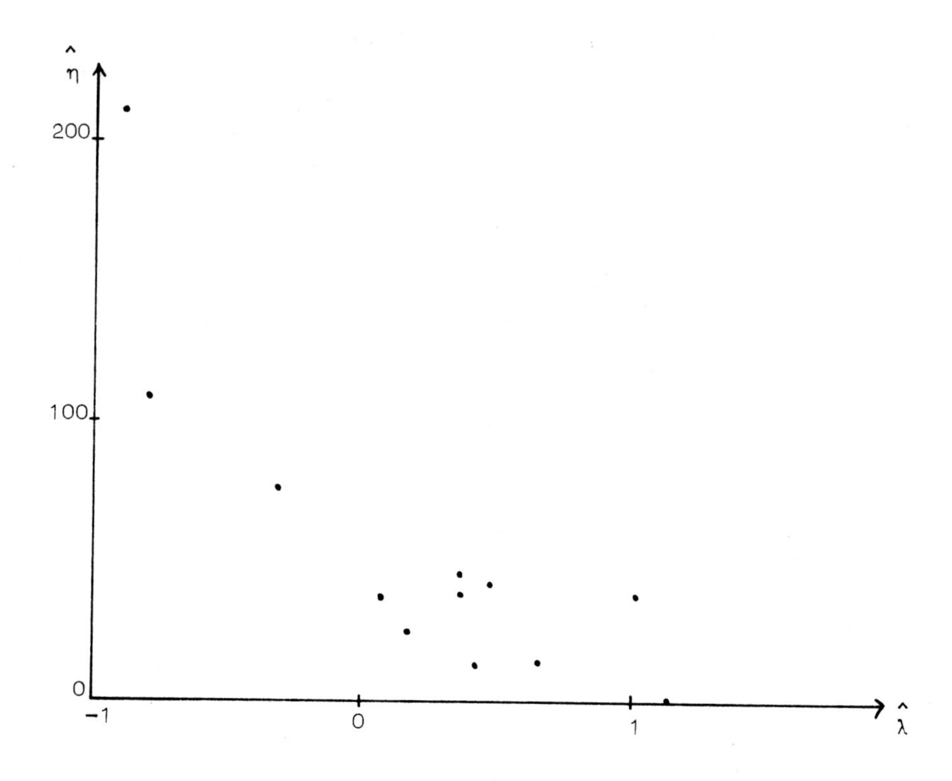

Figure 7.3. Plot of $\hat{\eta}^{(i)}$ against $\hat{\lambda}^{(i)}$.

A glance at Figure 4.1 shows that for the values of λ and ω in question here, $\hat{\omega}_\lambda$ varies very little with λ for a given sample. Hence the plot of $\hat{\omega}^{(i)}$ against $\hat{\lambda}^{(i)}$ (Figure 7.2) should show no systematic dependence between the two variables, and this seems roughly to be the case. However, the plot shows that the variation of $\hat{\lambda}^{(i)}$ is smaller for $\hat{\omega}^{(i)}$ small than for $\hat{\omega}^{(i)}$ large. This is not surprising because the generalized inverse Gaussian distribution becomes independent of λ for ω large (cf. (3.10)), so for large values of ω there is less information about λ than for small values. The same effect is also present in Figure 7.1. We conclude that Figure 7.2 indicates no disagreement with the hypothesis of a common value for λ.

The plot of $\hat{\eta}^{(i)}$ against $\hat{\lambda}^{(i)}$ (Figure 7.3) shows that $\hat{\eta}^{(i)}$ tends to decrease with $\hat{\lambda}^{(i)}$, but this effect can be explained. In fact equation (4.1a) shows that when $\hat{\omega}_\lambda$ does not vary too much, $\hat{\eta}_\lambda$ is roughly decreasing as a function of λ, because $R_\lambda(\omega)$ increases with λ, and Figure A.1 shows that the variation of R_λ is considerable for the values of ω in question.

Thus it seems quite reasonable to accept the hypothesis of a common value for λ and the next step is to determine which value λ might have.

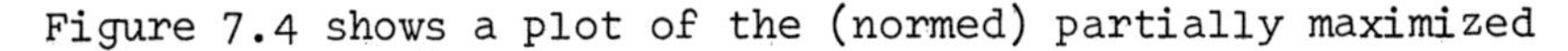

Figure 7.4 shows a plot of the (normed) partially maximized

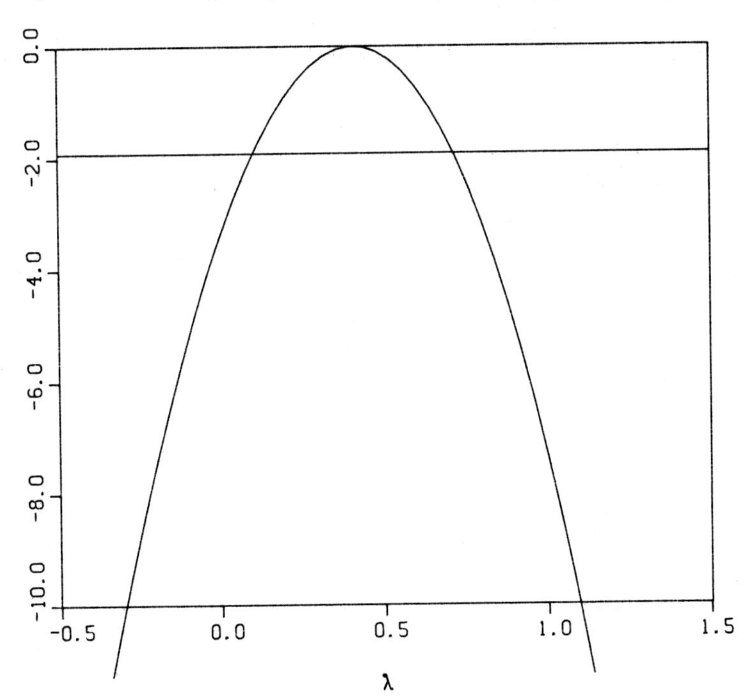

<u>Figure 7.4</u>. Plot of the partially maximized log-likelihood for λ when a common value for λ is assumed.

log-likelihood for λ, i.e. the sum of the 12 separate partially maximized log-likelihoods in Figure 7.1. The maximum of this concave function corresponds to the maximum likelihood estimate of λ, $\hat{\lambda} = 0.41$, and again the horizontal line indicates the approximate 95% confidence limits for λ based on the asymptotic $\chi^2(1)$ - distribution of $-2\ln Q$. The resulting confidence interval is $0.1 < \lambda < 0.75$, which in particular excludes the values $\lambda = 0$ and $\lambda = 1$. (For $\lambda = 1$ we have $-2\ln Q = 14.67$.)

It should be noted that if we repeat the analysis up to this point using only the 9 aircraft which have more than 11 observations we reach virtually the same conclusions.

It seems quite reasonable to analyze the data under the hypothesis $\lambda = 0.5$, a value which is rather close to the maximum likelihood estimate $\hat{\lambda} = 0.41$. This choice for λ is very convenient since the reciprocal observations then have an inverse Gaussian distribution. In fact it turns out that we can perform the inverse Gaussian analogue of an ordinary one-way analysis of variance (cf. section 5.4). Here $\eta^{(i)}$ and $\psi^{(i)-1}$ play respectively the role of the mean and the variance of the normal distribution. Table 7.4 shows the maximum likelihood estimates, which are given by $\hat{\eta}_{0.5}^{-1} = \bar{x}_{\dashv}$, $\hat{\psi}_{0.5}^{-1} = \bar{x}. - \bar{x}_{\dashv}^{-1}$.

Before carrying on, let us for a moment return to the question of a common value for λ. Figure 7.5 shows a plot of $\hat{\eta}_{0.5}^{(i)}$ against $\hat{\lambda}^{(i)}$ and again we look for unexpected systematic effects. There seem to be roughly no such effects, except perhaps that we see the same pattern as in Figure 7.2, i.e. the values of $\hat{\lambda}^{(i)}$ are more spread out for large values of $\hat{\eta}_{0.5}^{(i)}$ than for small ones,

Aircr. i	n_i	$\hat{\chi}_{0.5}^{(i)}$	$V(\hat{\chi}_{0.5}^{(i)})^{\frac{1}{2}}$	$\hat{\psi}_{0.5}^{(i)} \cdot 10^3$	$V(\hat{\psi}_{0.5}^{(i)})^{\frac{1}{2}} \cdot 10^3$	$\hat{\eta}_{0.5}^{(i)}$	$\hat{\omega}_{0.5}^{(i)}$	$\hat{\hat{\omega}}_{0.5}^{(i)}$	$V(\hat{\hat{\omega}}_{0.5}^{(i)})^{\frac{1}{2}}$
1	6	28.1	30.7	21.7	12.5	36.0	0.78	0.60	0.50
2	23	14.5	9.8	15.4	4.5	30.7	0.47	0.44	0.19
3	29	43.9	19.7	24.3	6.4	43.2	1.03	0.99	0.32
4	15	20.0	16.4	12.3	4.5	40.2	0.50	0.45	0.24
5	14	53.5	36.6	14.3	5.4	61.1	0.88	0.79	0.39
6	30	2.58	2.1	20.7	5.3	11.2	0.23	0.22	0.10
7	27	3.38	2.8	16.0	4.4	14.5	0.23	0.21	0.11
8	24	7.27	5.1	21.8	6.3	18.3	0.40	0.37	0.17
9	9	0.38	1.2	5.1	2.4	8.6	0.044	0.035	0.079
10	6	3.52	6.8	11.3	6.5	17.7	0.20	0.14	0.22
12	12	2.34	3.5	10.7	4.4	14.8	0.16	0.14	0.13
13	16	47.0	27.9	25.6	9.0	42.9	1.10	1.01	0.44

Table 7.4. Failures of airconditioning equipment. Maximum likelihood estimates of parameters for $\lambda = 0.5$ and their standard deviations.

but the clear pattern of Figure 7.3 has now been removed, as it should. We conclude that the plot provides no evidence against the hypothesis of a common value for λ.

From now on we assume that λ equals 0.5.

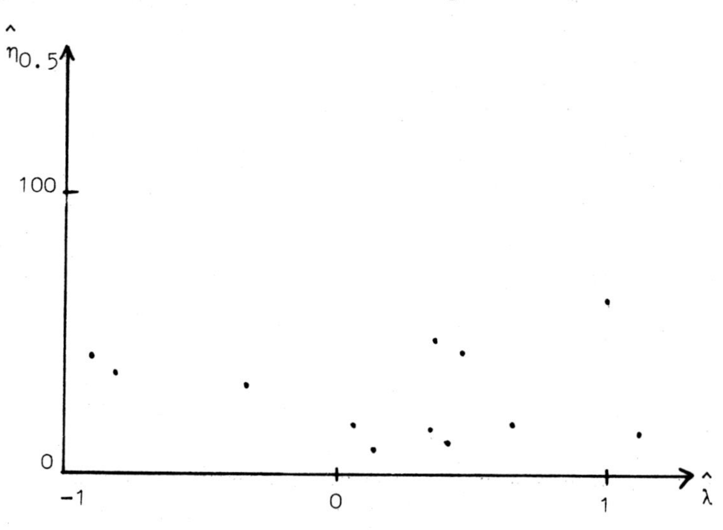

Figure 7.5. Plot of $\hat{\eta}^{(i)}_{0.5}$ against $\hat{\lambda}^{(i)}$.

From the estimates of $\psi^{(i)}$ and their variances (Table 7.4) we are led to ask whether there is a common value for ψ (this is the analogue of the hypothesis of homogeneity of variances). For the i'th aircraft we have

$$n_i \psi^{(i)} \hat{\psi}^{(i)^{-1}}_{0.5} \sim \chi^2(n_i - 1),$$

where n_i is the number of observations for the aircraft. Hence we can test equality of the $\psi^{(i)}$'s by the usual Bartlett-statistic ($B = 16.53$, $\chi^2(11)$, $p = 10\%$), which shows no significance. Again we make a graphical check of the hypothesis, this time by plotting

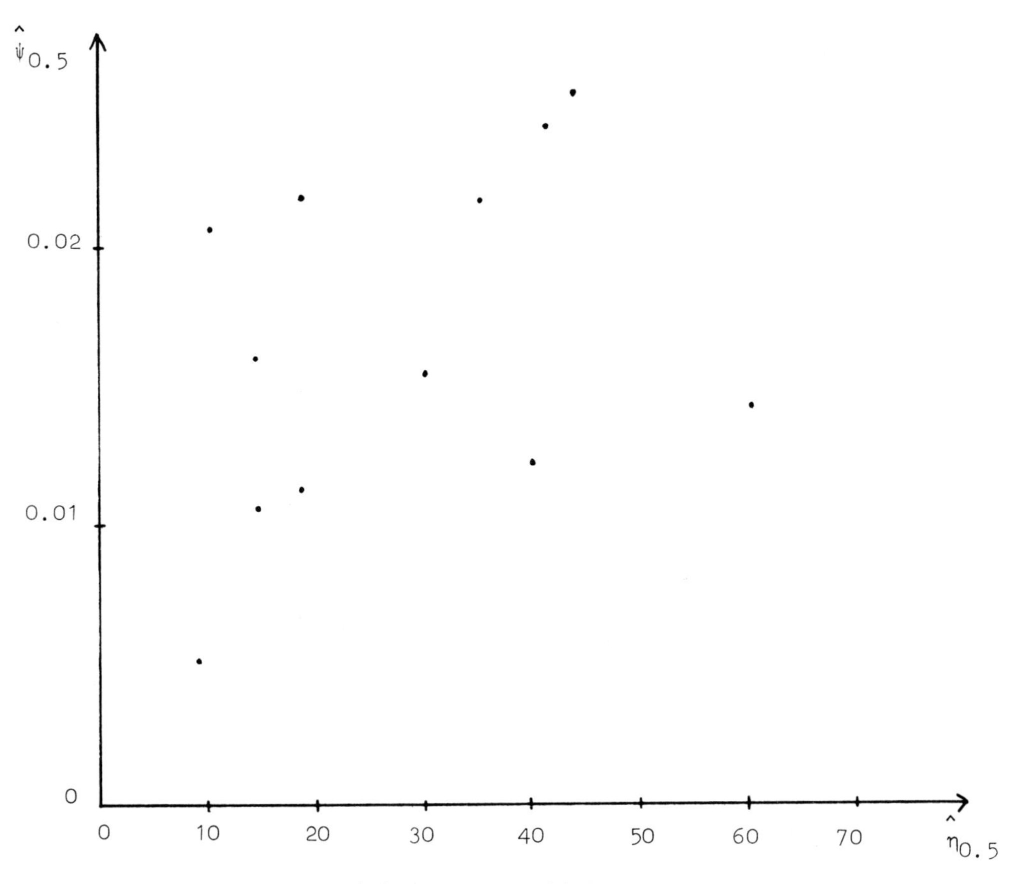

Figure 7.6. Plot of $\hat{\psi}_{0.5}^{(i)}$ against $\hat{\eta}_{0.5}^{(i)}$.

$\hat{\psi}_{0.5}^{(i)}$ against $\hat{\eta}_{0.5}^{(i)}$ (Figure 7.6). One notes that aircraft 9 corresponds to the lower left point in Figure 7.6, and it is clear from for example the estimates in Table 7.2 that this aircraft is somewhat extreme compared to the remaining ones. Otherwise the plot does not indicate any departures from the hypothesis.

The maximum likelihood estimate for the common value of ψ is 0.014, whereas the estimates for $\eta^{(i)}$ are unchanged under the hypothesis of a common value for ψ.

If we accept the hypothesis of a common value for ψ we can test the hypothesis of a common value for η by an F-test (cf. Chhikara and Folks, 1978) ($F = 2.21$, (11, 199) degrees of freedom, $p < 2.5\%$). The test leaves no doubt that the $\eta^{(i)}$'s are different and we conclude that the distribution of intervals between failures is not the same for all aircraft.

We shall now examine, for each aircraft, the marginal likelihood for ω (still assuming $\lambda = 0.5$), obtained from the marginal distribution of the resultant length. As discussed in section 5.3 we use the marginal distribution of the resultant length when we want to test hypotheses about ω, and we shall here test the hypothesis $\omega = 0$, which corresponds to the intervals between failures having a gamma distribution with shape parameter $\lambda = 0.5$. We also consider procedures for setting up approximate confidence intervals for ω.

Figure 7.7 shows two typical examples of marginal log-likelihoods for ω. As noted in section 5.3 the marginal log-likelihood has a finite value for $\omega = 0$ and has an approximately linear right tail. The two log-likelihoods appear to be concave or, at least, unimodal, in concordance with Theorem 5.4. Recall here that the conjecture in Theorem 5.4 has in fact been proved in the case $\lambda = 0.5$, cf. page 88.

The horizontal line in the plots indicates the 95% confidence intervals for ω based on a $\chi^2(1)$-distribution for $-2\ln Q$, testing that ω has some given value. It is not known whether $-2\ln Q$ in fact has a limiting (or exact) χ^2-distribution in this case, though it seems probable for $\omega > 0$ (cf. the remark

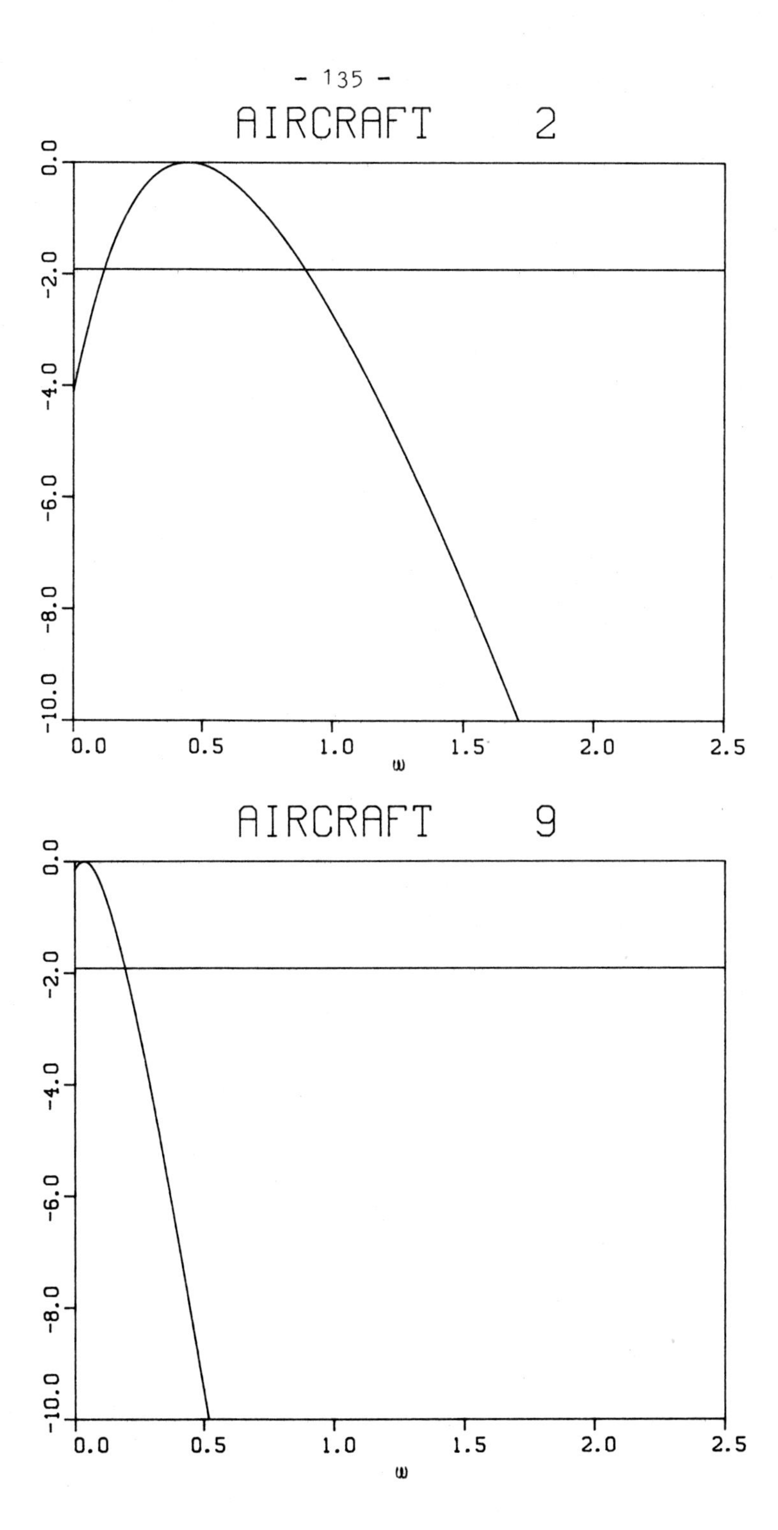

Figure 7.7. Typical examples of marginal log-likelihoods for ω.

right after Theorem 5.4), but the line in the plots should at least give an idea of the precision of the maximum likelihood estimate.

The log-likelihood for aircraft 2 in Figure 7.7 shows a moderate skewness corresponding to a value $|F_\omega| = 0.54$ of Sprott's measure for the deviation of the likelihood from normality (cf. (5.27)). In fact we also have $|F_\omega| = 0.54$ for aircraft 9 in the figure. We have tried to find a transformation $\nu = \nu(\omega)$ of the parameter which improves the normality of the likelihood. Sprott (1973) and Efron and Hinkley (1978) consider such transformations, but in the present case it is hardly feasible to write down analytical expressions for these transformations.

Instead we have examined the effect of a number of simple transformations and we have found that the following transformation

$$\nu(\omega) = \omega/\sqrt{\omega + 0.5} \qquad (7.1)$$

yields good results for the present data. The transformation (7.1) almost removes the skewness of the likelihood, considerably improves the normality of the likelihood and causes a systematic decrease in the value of Sprott's $|F|$.

Figure 7.8 shows the log-likelihood as a function of ν for the two aircraft which we considered above. The dashed curve in the plots is the approximating normal log-likelihood having the correct second derivative at the maximum, as given by the observed information $\hat{j}_\lambda$ in (5.26). The log-likelihood for aircraft 2 in Figure 7.8 is a typical example where the normal approximation

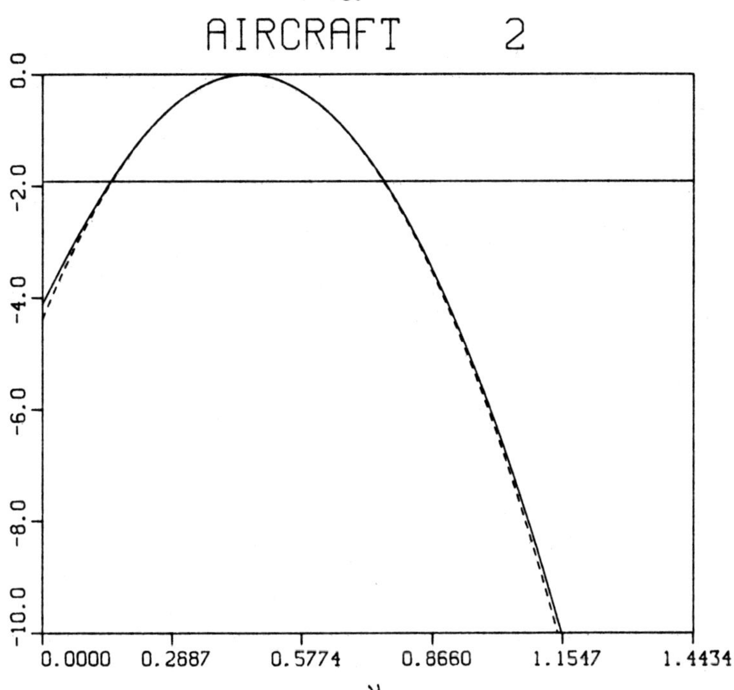

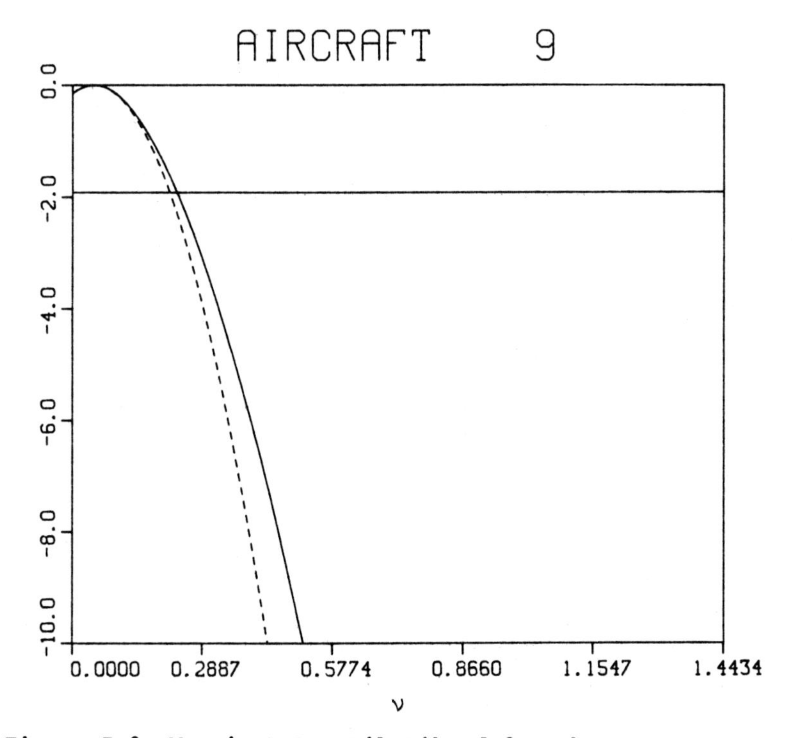

Figure 7.8. Marginal log-likelihood for the parameter $\nu = \omega/\sqrt{\omega + 0.5}$ (solid curve) and the approximating normal log-likelihood (dashed curve).

is very good in the vicinity of the maximum, and the new value for Sprott's $|F|$ is 0.43. The second plot in Figure 7.8 (aircraft 9) shows an example where the approximation is not as good as in the average case, and Sprott's $|F|$ has only been reduced to the value 0.51. The reason for this appears to be a small maximum-likelihood estimate for ω in combination with a moderate sample size $(n = 9)$.

The normalizing transform (7.1) should probably be different for other values of λ, but a suitable general form is not known. However, in section 7.3 we consider an example where a reasonable normalizing transform for $\lambda = 0$ is $\nu(\omega) = \sqrt{\omega}$, suggesting that the general form could be $\nu_\lambda(\omega) = \omega/\sqrt{\omega + |\lambda|}$, but the subject needs further investigation.

Table 7.5 shows approximate 95% confidence intervals for ω obtained by treating the variable

Aircraft	n	Confidence interval for ω	$(\bar{x}.\bar{x}_{\dashv}-1)/(n-1)$	p
1	6	0.00 - 2.05	0.26	10.0 %
2	23	0.12 - 0.89	0.096	0.5 %
3	29	0.45 - 1.72	0.035	< 0.05%
4	15	0.10 - 1.04	0.14	2.5 %
5	14	0.18 - 1.77	0.088	0.5 %
6	30	0.04 - 0.45	0.15	2.0 %
7	27	0.03 - 0.46	0.17	2.5 %
8	24	0.09 - 0.76	0.11	0.5 %
9	9	0.00 - 0.18	2.86	90.0 %
10	6	0.00 - 0.63	1.00	60.0 %
12	12	0.00 - 0.43	0.57	30.0 %
13	16	0.30 - 2.09	0.061	0.1 %

Table 7.5. The approximate 95% confidence intervals for ω, values of the test statistic $(\bar{x}.\bar{x}_{\dashv}-1)/(n-1)$ and the corresponding p-values.

$$\sqrt{\hat{j}_{0.5}(\frac{d\omega}{d\nu})^2}\ (\hat{\nu}_{0.5} - \nu) \tag{7.2}$$

as a standard normal variate, where $\hat{j}_{0.5}(\frac{d\omega}{d\nu})^2$ is the observed information for ν, according to (5.26). Note that for aircraft 2 these confidence intervals are virtually equivalent to likelihood intervals, and in fact this seems often to be the case. It is not known whether (7.2) in fact has a limiting normal distribution (cf. the remark right after Theorem 5.4), but for the present data the difference between the observed information $\hat{j}_{0.5}$ and the expected information $ni_{0.5}(\hat{\hat{\omega}}_{0.5})$ (cf. (5.32)) is negligible. Hence, in line with Efron and Hinkley (1978) we prefer in (7.2) to use the observed information, provided, of course, that it can be proved that (7.2) has an asymptotic standard normal distribution.

In cases where the left endpoint of the confidence interval based on (7.2) was negative we have replaced it by a zero, but in such cases the right endpoint of the interval probably does not give a very good approximation.

Finally, Table 7.5 gives the values of the test statistic $(\bar{x}.\bar{x}_{-1} - 1)/(n-1)$, which for $\omega = 0$ and $\lambda = 0.5$ is distributed as $F_{n-1,1}$ (cf. (5.8)). This statistic gives the exact test for $\omega = 0$, small values being significant. The p-values in the table show departures from the hypothesis at the 5% level for all aircraft, except for number 1, 9, 10 and 12.

One recalls that the test for $\omega = 0$, assuming $\lambda = 1$, (the test for exponentiality) made at an earlier stage of the analysis showed departures only for aircraft 3 and 13, and hence the test for $\omega = 0$ based on the resultant length appears to be sensitive to the value of λ for which it is performed. It should also be recalled that the later stages of the analysis showed that the value $\lambda = 1$ is quite improbable. One may say that if we consider the test based on the resultant length as a test for the gamma distribution against generalized inverse Gaussian alternatives, this procedure lacks "inference robustness" in the sense of Box and Tiao (1973, pp. 152-153), because the conclusions to be drawn depend on whether the test is performed for the correct value of λ.

The above test is, of course, only an approximation to the exact test for the gamma distribution against generalized inverse Gaussian alternatives, which is based on the conditional distribution of $X_{\dashv}$ given $(X_{\cdot}, X_{\sim})$.

To judge the fit of the generalized inverse Gaussian distribution we have plotted, for each aircraft, the empirical log-survivor function, i.e. the points $(x_{(j)}, \ln(1-(j-\frac{1}{2})/n))$, $j = 1,\ldots,n$, where $x_{(j)}$ is the j'th order statistic, and the estimated log-survivor function, i.e. the theoretical log-survivor function with $\lambda^{(i)} = 0.5$, $\chi^{(i)} = \hat{\chi}^{(i)}_{0.5}$ and $\psi^{(i)} = \hat{\psi}^{(i)}_{0.5}$ (Figure 7.9). We have also made the plots using the common value for ψ (these plots are not shown), but this gave some discrepancies in the tail of the log-survivor function. (Recall that ψ determines the slope of the tail of the log-survivor function.) Also shown in Figure 7.9 are the plots corresponding to the reciprocal observations, which are interesting since the reciprocal observations should also follow a generalized inverse Gaussian distribution.

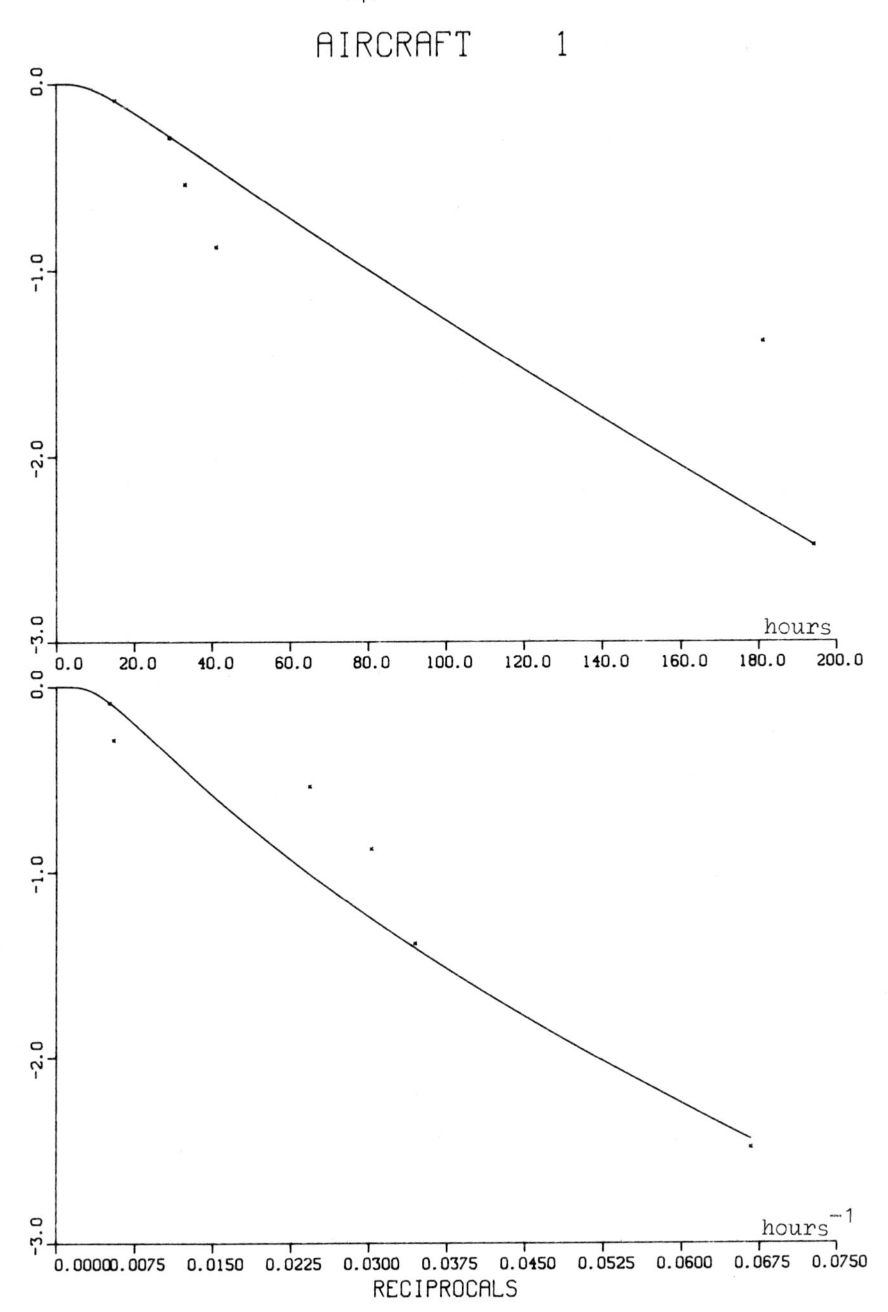

Figure 7.9. Empirical and estimated log-survivor functions for untransformed observations (top) and reciprocals (bottom).

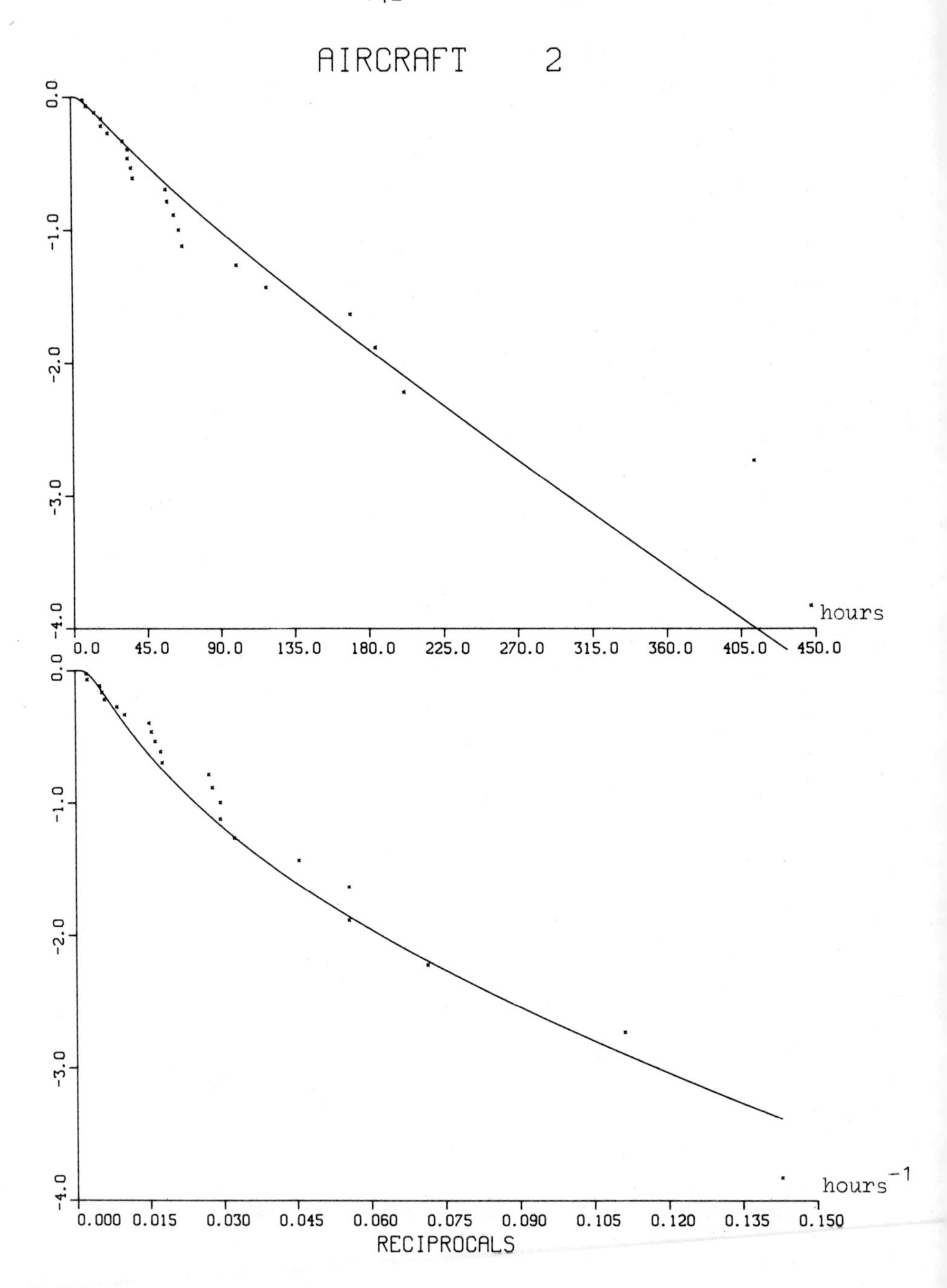

Figure 7.9. (Continued.)

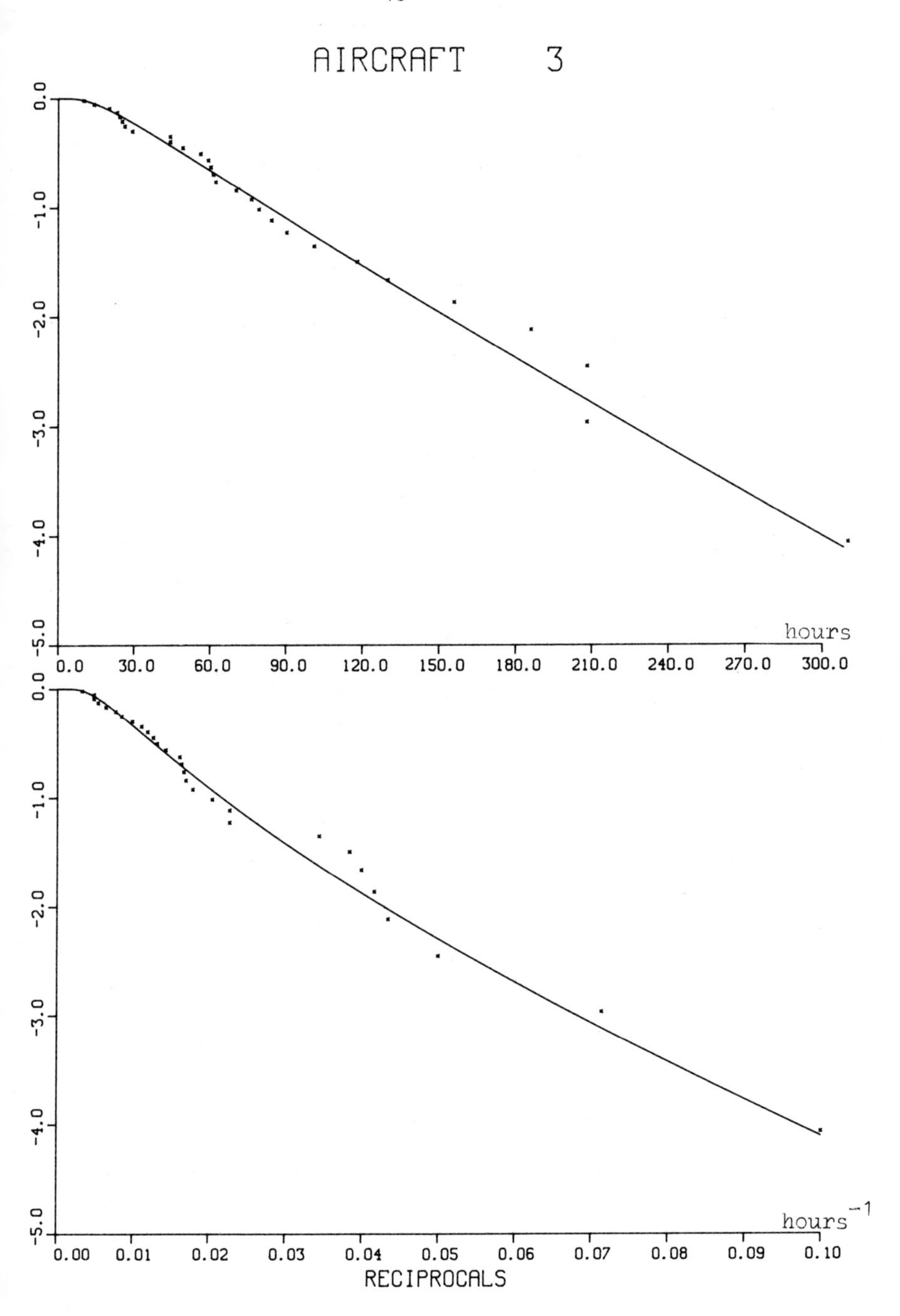

Figure 7.9. (Continued.)

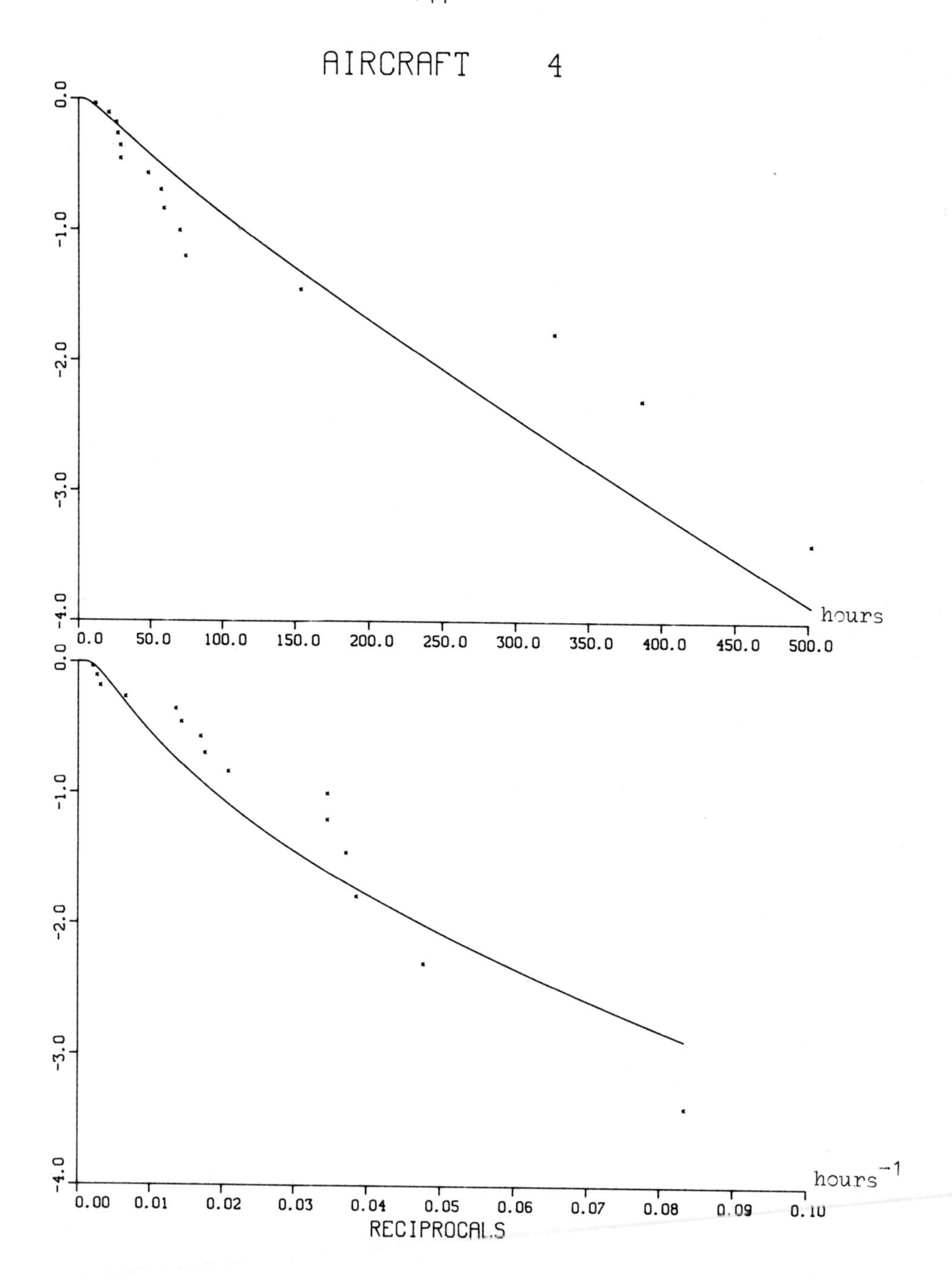

Figure 7.9. (Continued.)

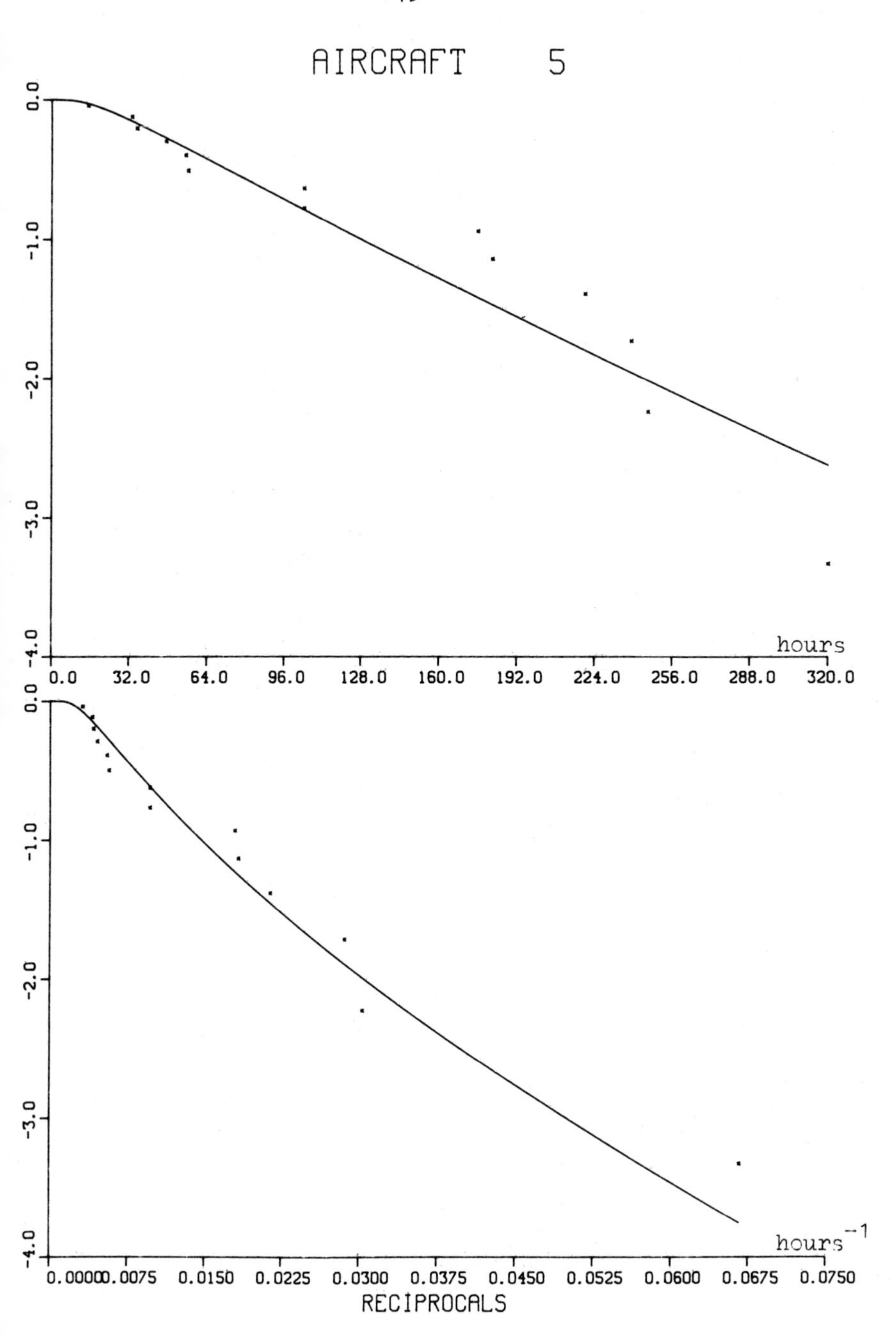

Figure 7.9. (Continued.)

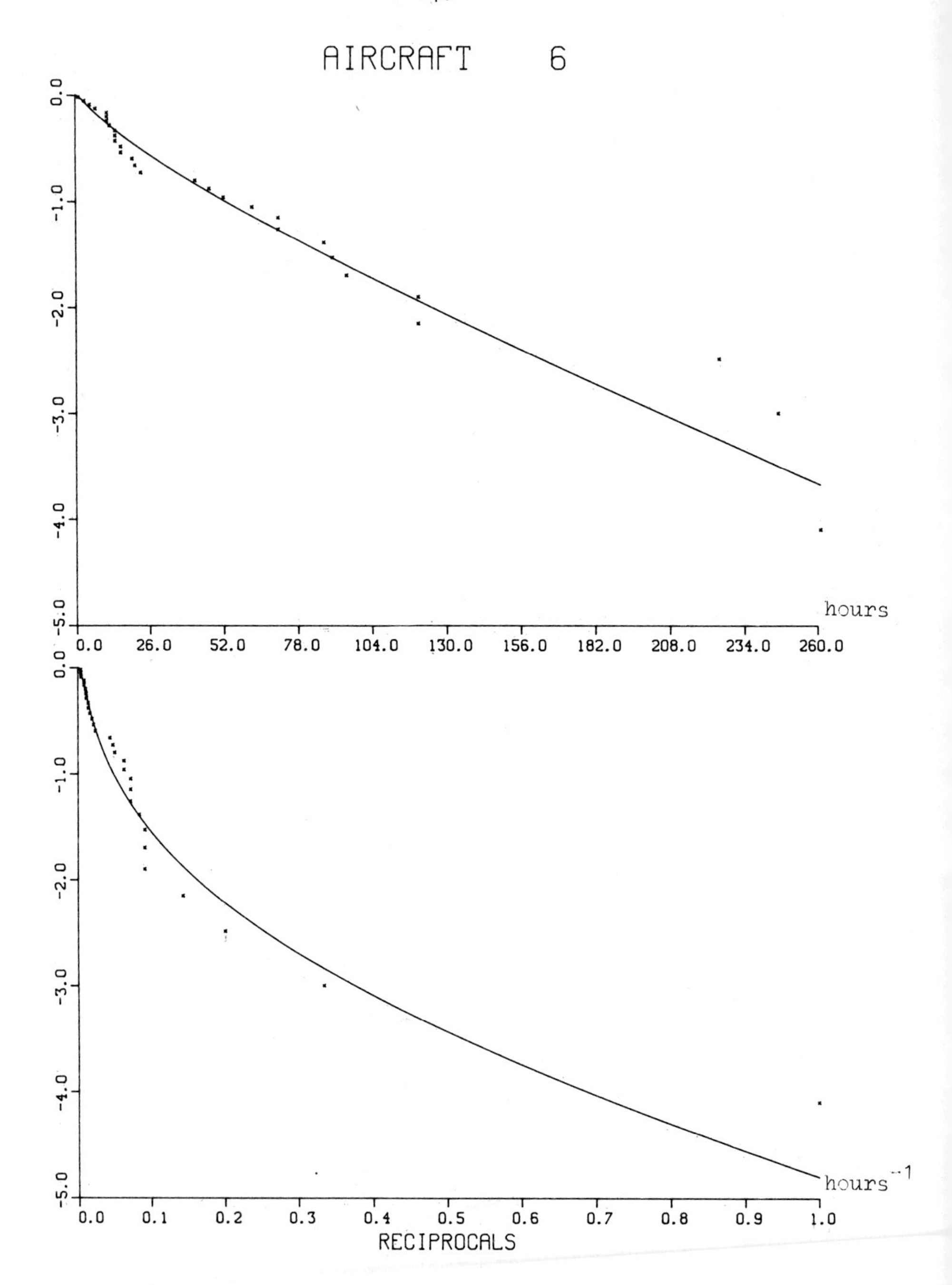

Figure 7.9. (Continued.)

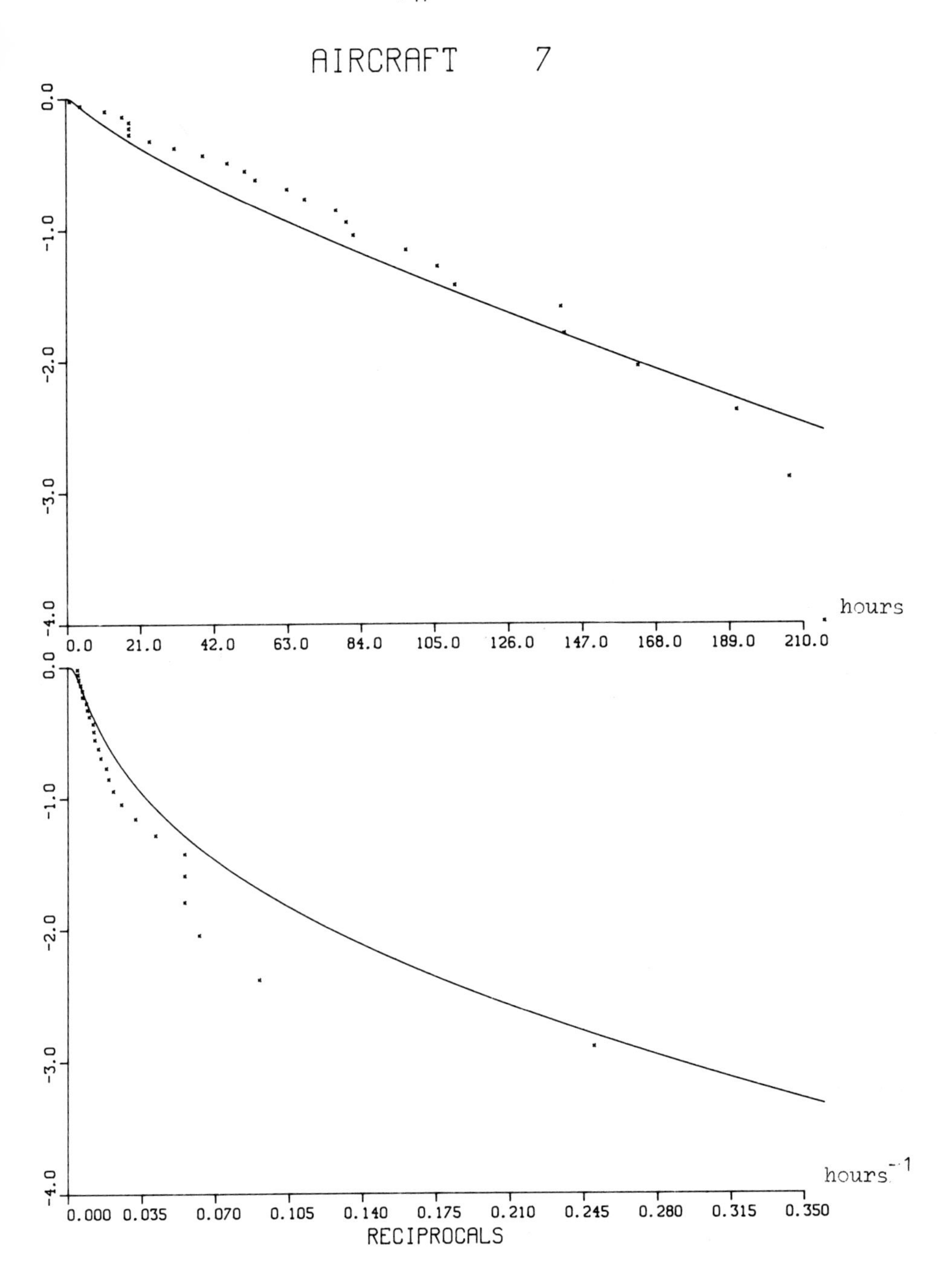

Figure 7.9. (Continued.)

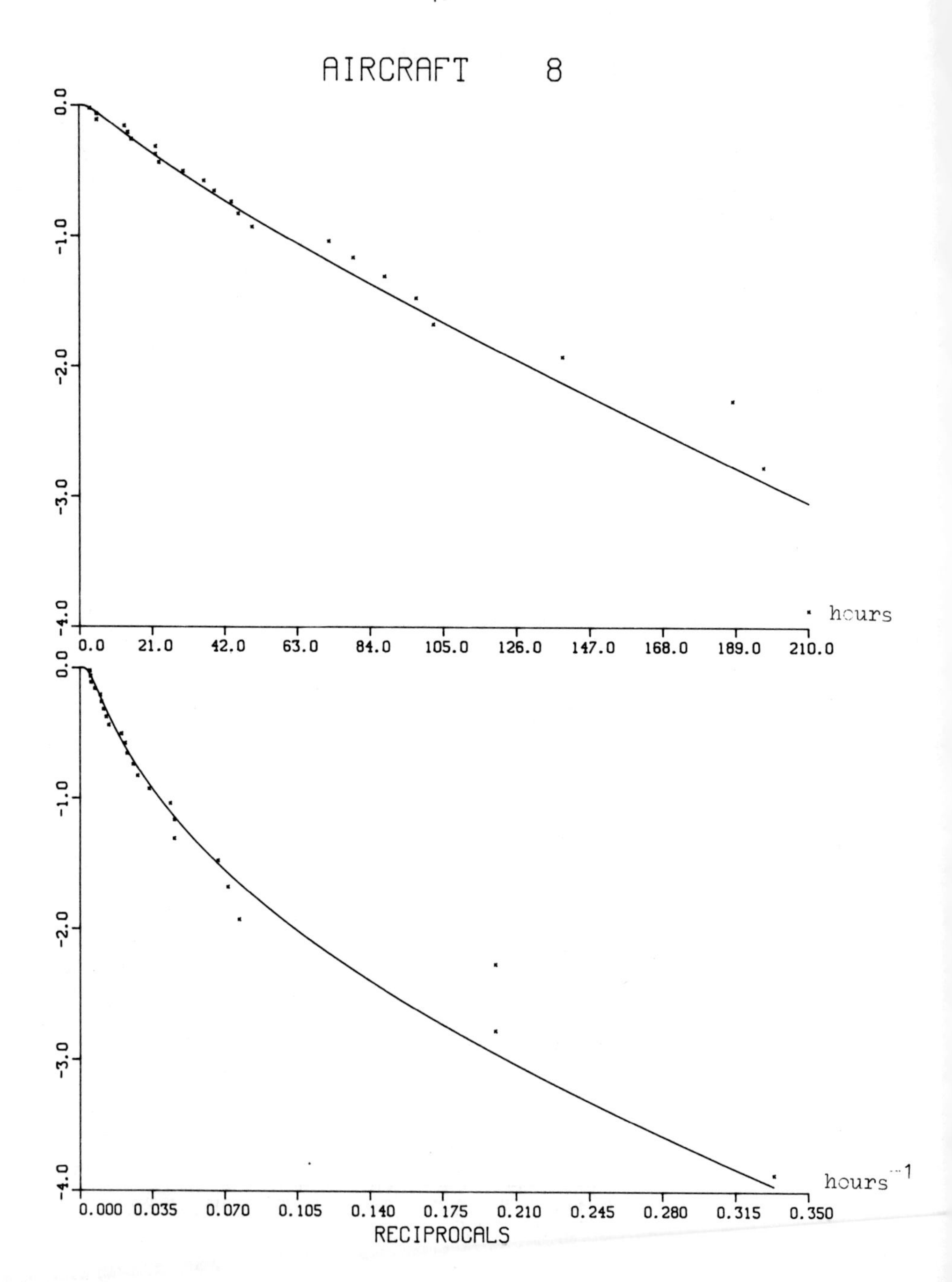

Figure 7.9. (Continued.)

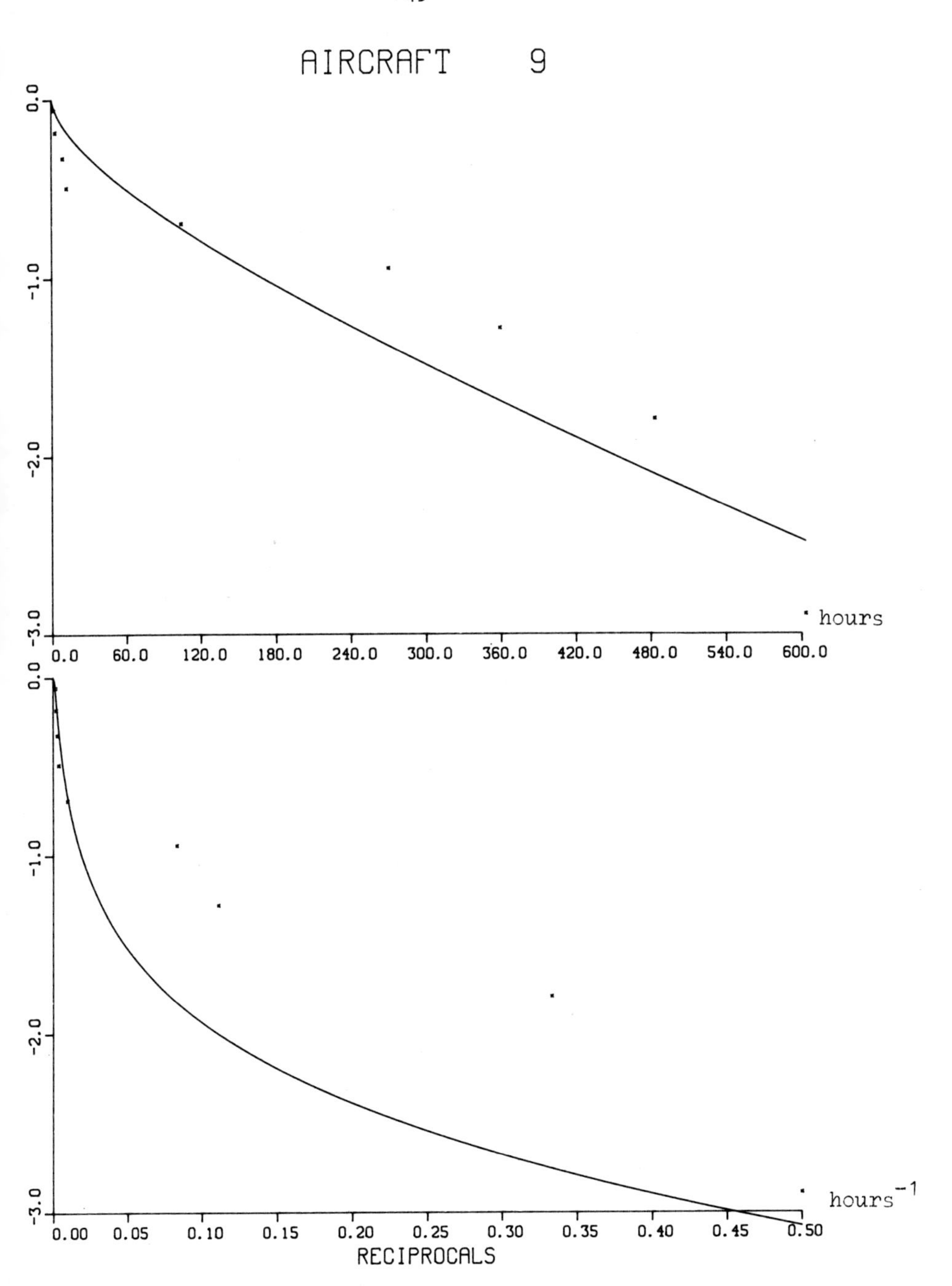

Figure 7.9. (Continued.)

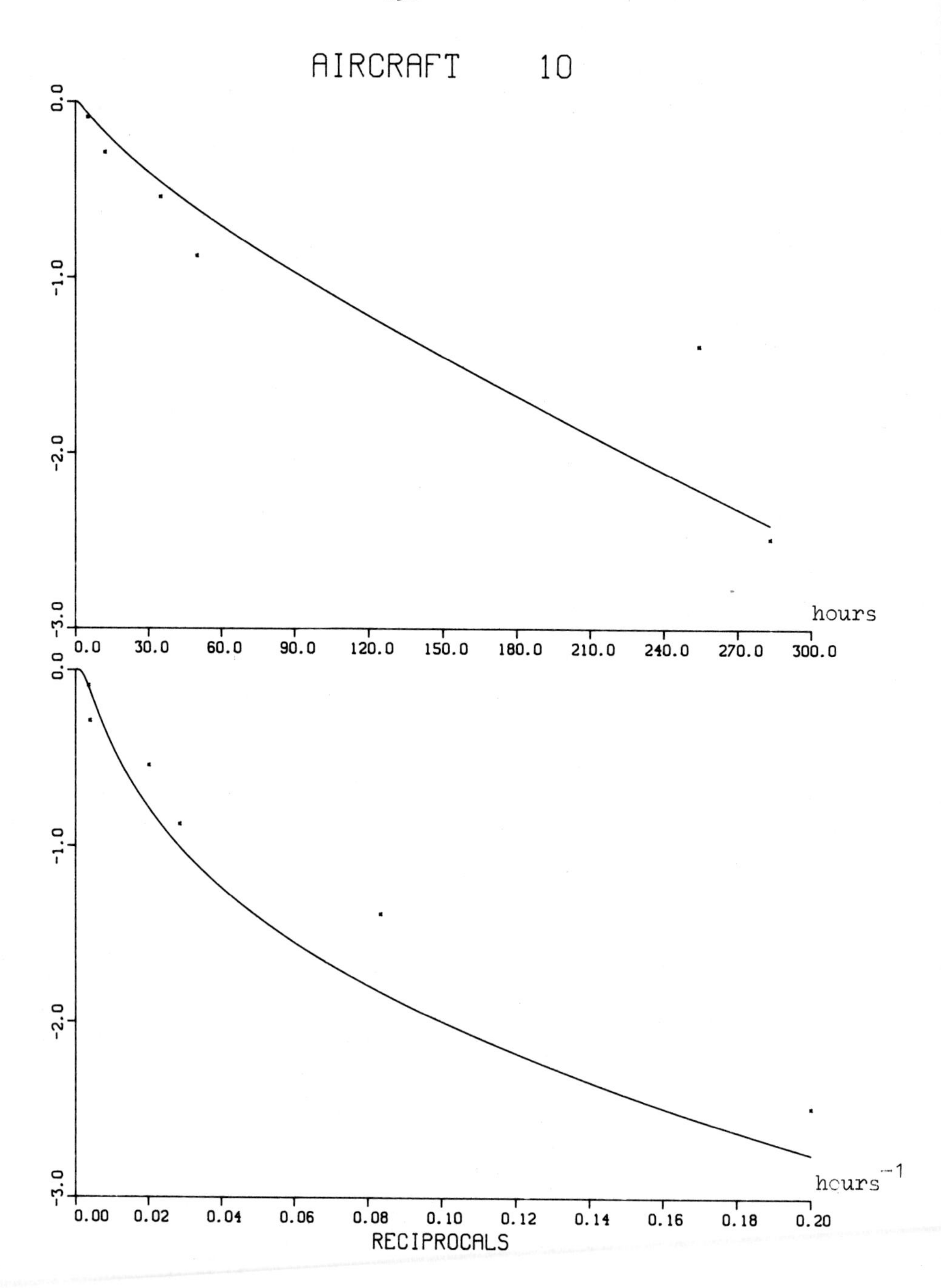

Figure 7.9. (Continued.)

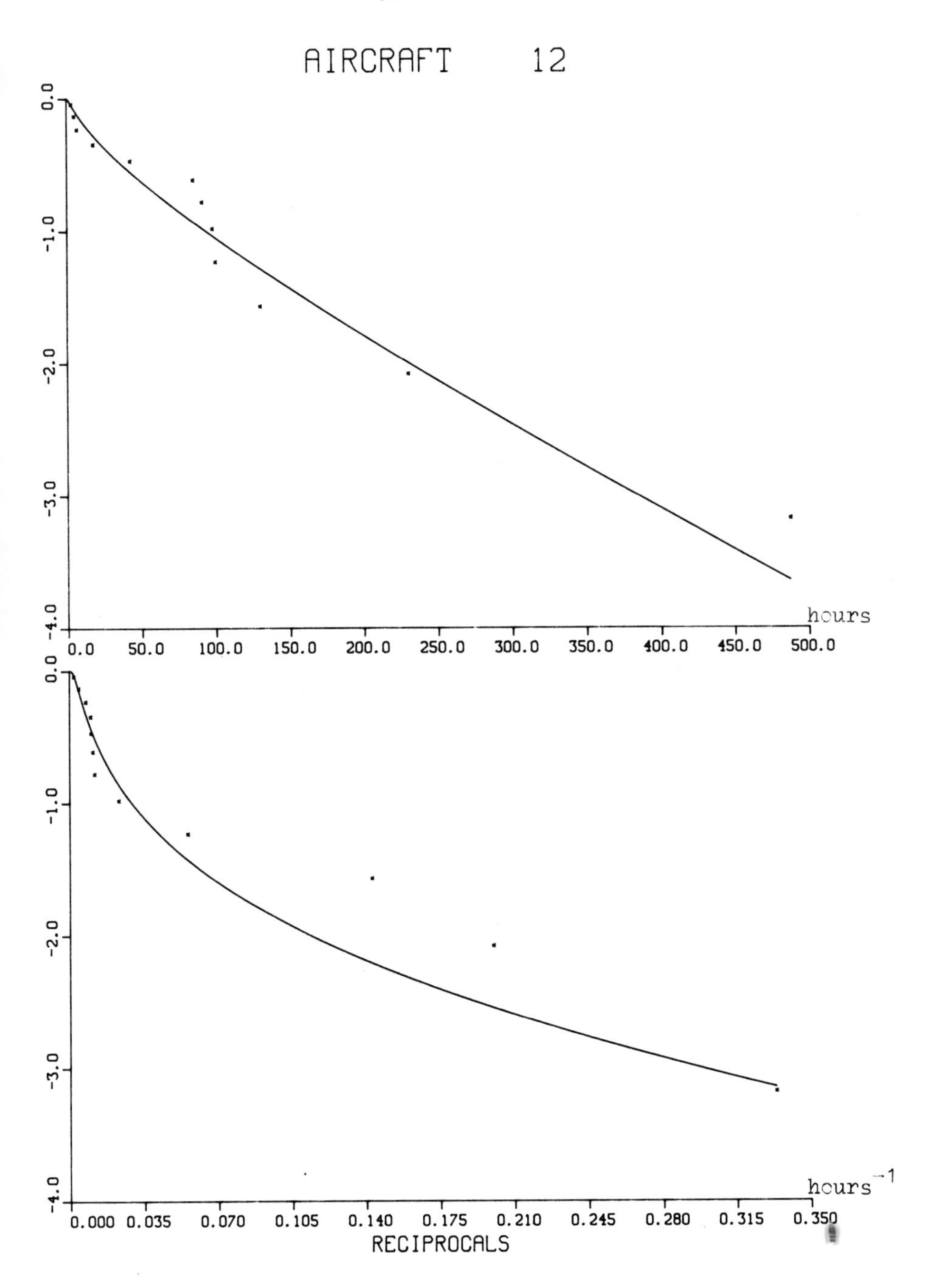

Figure 7.9. (Continued.)

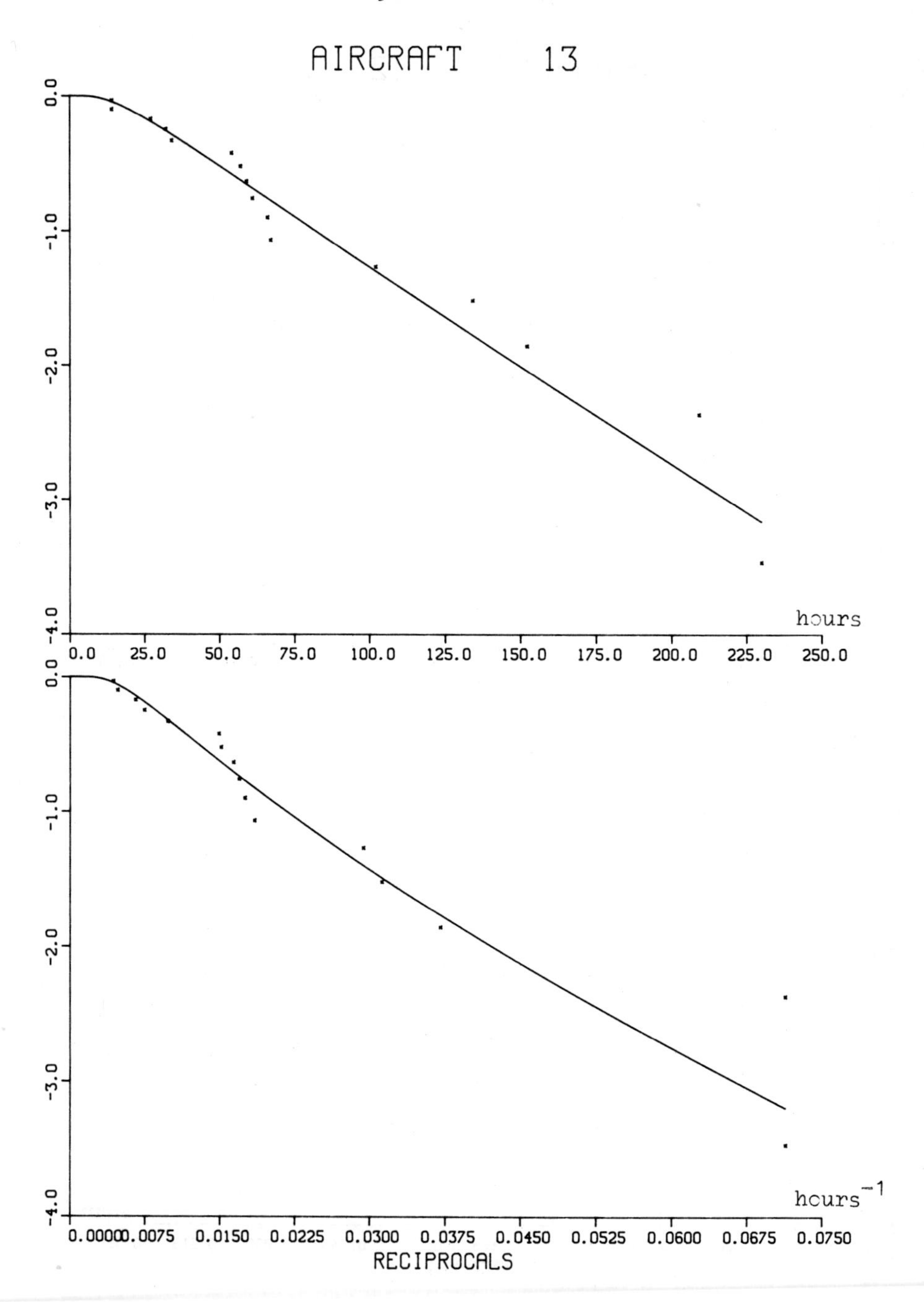

Figure 7.9. (Continued.)

The fit appears to be quite reasonable, except perhaps for aircraft 7 for which the distribution is not far from being exponential, since the empirical log-survivor function is almost linear (this was confirmed by our first test for exponentiality). For the remaining aircraft Figure 7.9 shows that the estimated log-survivor function has an initial departure from linearity, but quickly approaches the linear tail.

The values of the Kolmogorov-Smirnov test statistic (Table 7.6) confirm that the fit is quite good. We conclude that with the above exception (aircraft 7) we have found clear discrepancies from the exponential distribution, whereas for the remaining 11 aircraft the generalized inverse Gaussian distribution with $\lambda = 0.5$ provides a good fit to the data.

Aircraft	D_n	n	$\sqrt{n}\,D_n$
1	0.304	6	0.744
2	0.165	23	0.793
3	0.0793	29	0.427
4	0.254	15	0.984
5	0.185	14	0.694
6	0.138	30	0.756
7	0.145	27	0.753
8	0.0725	24	0.355
9	0.283	9	0.849
10	0.223	6	0.546
12	0.188	12	0.650
13	0.149	16	0.596

Table 7.6. Values of the Kolmogorov-Smirnov test statistic D_n, for the fitted generalized inverse Gaussian distribution with $\lambda = 0.5$. (The upper 5% point in the asymptotic distribution of $\sqrt{n}\,D_n$ is 1.358, assuming a known distribution function, see Cox and Lewis (1966).)

7.2 Pulses along a nerve fibre

Table 7.7 shows a record of 799 intervals between pulses along a nerve fibre. As noted by Cox and Lewis (1966), the empirical log-survivor function for these data (Figure 7.10) is almost linear, so the distribution is not far from being exponential. This is roughly confirmed by the estimates of the parameters of the generalized inverse Gaussian distribution, which are

$$\hat{\lambda} = 0.804, \quad \hat{\chi} = 0.919, \quad \hat{\psi} = 0.171.$$

However, as Cox and Lewis also found, there is a certain discrepancy from exponentiality.

The partially maximized log-likelihood for λ (Figure 7.11) shows that the value $\lambda = 1$ is near the right endpoint of the 95% confidence interval for λ, in fact $-2\ln Q = 3.73$ for $\lambda = 1$. But anyway we shall see that ω is certainly not zero. The estimates for ω in the marginal distribution of the resultant length in the cases $\lambda = \hat{\lambda}$ and $\lambda = 1$ are $\hat{\hat{\omega}}_{\hat{\lambda}} = 0.395$ and $\hat{\hat{\omega}}_1 = 0.320$, and if we standardize according to (5.32) we get

$$\sqrt{n i_{\hat{\lambda}}(\hat{\hat{\omega}}_{\hat{\lambda}})}\ \hat{\hat{\omega}}_{\hat{\lambda}} = 10.1, \qquad \sqrt{n i_1(\hat{\hat{\omega}}_1)}\ \hat{\hat{\omega}}_1 = 7.0,$$

showing a clear deviation from the hypothesis $\omega = 0$ for both values of λ.

1·0	2·5	5·5	13·5	6·0	2·5	18·0	4·0	0·5
1·5	13·0	3·5	14·0	5·0	13·5	2·5	35·5	13·5
3·0	11·5	22·0	7·5	1·5	12·5	0·5	2·0	9·5
6·0	10·0	6·5	13·0	5·5	5·0	22·0	5·5	1·0
15·5	38·0	2·0	0·5	10·0	8·0	7·0	6·5	10·5
32·0	2·5	19·5	3·0	8·0	4·0	7·0	8·5	2·0
4·0	1·0	7·0	17·5	5·0	26·0	7·0	17·0	5·0
7·5	0·5	5·0	1·5	1·5	22·0	4·0	11·5	3·0
3·0	10·0	4·0	13·0	11·5	9·5	25·5	4·0	24·0
3·0	10·5	1·0	2·5	6·0	36·0	9·0	13·0	6·0
7·5	1·0	28·5	9·0	0·5	6	1·0	1·5	7·5
34·0	2·0	17·5	23·0	6·0	15·0	25·5	10·5	6·0
15·0	8·0	8·5	6·0	8·5	7·0	3·0	22·5	26·0
1·0	16·0	10·5	11·5	7·0	22·5	11·0	20·0	24·0
2·0	21·5	7·0	16·0	5·0	21·0	0·5	1·5	14·5
1·0	10·0	38·5	4·0	1·0	4·5	4·5	8·0	28·5
40·5	6·5	3·0	13·0	6·5	3·5	11·0	3·0	11·0
4·5	5·0	17·0	41·0	3·0	31·0	29·5	14·5	0·5
9·5	10·0	7·5	5·0	10·5	25·5	1·5	21·5	22·0
7·0	4·0	14·5	34·5	25·0	25·0	35·5	1·5	2·5
6·0	40·5	4·0	7·5	2·0	23·5	7·0	5·0	24·5
18·0	5·5	36·0	0·5	21·0	14·0	1·0	5·0	5·0
1·0	4·5	15·5	19·5	14·5	2·0	25·5	15·5	9·5
5·5	13·0	10·0	2·0	4·0	33·0	1·5	13·5	22·0
2·0	7·5	5·0	6·5	0·5	4·0	20·5	13·5	1·0
4·0	18·0	0·5	17·0	15·0	5·5	8·5	16·5	36·0
8·5	9·0	12·0	6·5	22·5	1·5	18·5	7·0	4·5
2·0	5·0	3·5	6·5	3·0	16·0	19·5	4·5	2·0
2·5	17·0	11·0	15·0	12·5	8·0	41·0	13·5	1·0
7·0	28·0	24·5	14·5	1·0	5·5	40·5	7·0	1·0
3·5	4·5	1·5	11·5	3·0	13·0	12·0	4·5	3·0
19·5	7·5	9·0	0·5	1·0	2·5	26·0	4·0	11·0
6·5	7·0	23·5	19·0	8·5	3·5	20·0	3·0	26·5
28·0	7·5	18·5	2·0	5·0	2·0	12·0	8·0	9·0
6·0	11·0	8·5	4·0	14·0	11·0	3·0	1·0	5·0
15·5	16·5	21·0	7·5	10·5	4·0	36·5	3·5	5·0
2·5	2·0	1·0	5·0	14·0	4·0	13·5	9·5	1·5
5·0	3·5	11·0	31·0	15·0	0·5	9·0	5·5	4·0
6·5	4·5	6·0	41·5	1·0	3·0	0·5	5·0	7·5
2·5	9·0	0·5	5·5	1·0	2·5	8·5	8·5	2·5
0·5	4·0	17·0	35·5	14·0	2·5	1·0	12·0	6·5
4·5	3·5	20·5	4·0	4·5	8·0	5·5	0·5	1·0
1·5	3·5	13·5	30·5	35·5	2·5	13·0	6·5	5·0
13·5	34·0	3·5	9·0	3·0	6·5	6·5	10·5	25·5
8·5	13·5	15·0	2·5	6·0	21·0	34·0	1·5	
1·5	10·5	4·5	10·0	14·5	10·5	6·5	19·5	

10·5	25·5	35·0	3·5	0·5	17·0	8·0	8·5	10·5
1·5	5·5	2·0	46·5	13·5	8·0	2·5	19·0	1·0
2·5	14·0	4·0	2·0	2·5	32·5	17·0	14·0	31·5
5·5	18·0	8·0	48·0	1·5	2·0	5·0	7·0	17·5
29·5	7·0	19·0	7·0	20·0	33·5	1·0	3·5	0·5
3·0	27·5	4·0	69·0	2·0	5·0	2·0	7·0	19·0
9·0	14·0	16·0	7·5	10·5	25·5	5·0	1·5	21·5
27·5	2·0	19·5	0·5	14·5	13·0	11·0	10·5	1·5
18·5	0·5	29·0	2·5	12·0	3·5	12·0	20·0	19·5
4·5	47·0	28·0	11·5	4·0	35·5	2·0	2·0	2·0
7·0	36·5	37·0	15·5	11·5	5·5	14·0	5·5	8·5
9·5	2·5	7·5	2·5	5·0	23·5	5·0	22·0	11·5
1·0	3·5	3·5	2·5	9·5	1·0	11·5	45·0	39·0
7·0	5·5	13·0	14·5	10·0	19·0	1·5	5·0	7·0
4·5	19·0	12·5	0·5	13·0	2·0	17·0	24·5	4·0
2·5	10·5	0·5	37·0	3·0	21·5	10·5	4·5	5·5
7·5	24·5	8·5	15·0	20·0	5·5	20·5	0·5	3·5
11·5	19·0	32·0	4·5	25·5	11·5	7·5	4·0	22·5
7·5	19·0	30·5	1·0	7·5	7·0	2·5	3·0	23·0
4·0	0·5	7·5	9·5	55·0	4·0	8·5	4·0	10·0
12·0	3·0	13·0	23·5	8·0	56·0	26·5	0·5	9·5
8·0	6·5	1·5	0·5	39·0	25·0	15.0	7·5	25·0
3·0	3·0	2·5	25·5	2·0	12·5	7·5	25·0	4·5
5·5	0·5	17·0	6·0	13·5	9·0	9·5	18·0	11·0
7·5	8·0	3·5	6·0	17·5	6·0	3·5	4·0	14·5
4·5	2·5	5·0	21·5	35·5	1·0	41·5	17·0	0·5
1·5	5·0	4·5	16·0	7·5	7·5	2·0	1·0	9·5
10·5	8·0	1·0	4·5	14·5	6·0	2·0	10·5	3·0
1·0	3·0	15·0	10·0	2·0	4·0	7·0	16·0	19·5
7·0	3·0	3·5	1·5	0·5	19·0	17·0	11·0	4·0
12·0	3·0	6·0	2·5	14·0	11·0	5·0	25·5	1·5
14·5	3·0	0·5	6·5	10·5	8·0	7·5	6·0	14·0
8·0	5·5	8·0	7·5	4·5	20·0	2·5	8·0	4·5
3·5	22·0	7·0	2·5	8·5	29·0	2·0	26·0	8·5
3·5	2·5	24·5	4·0	4·5	2·5	2·5	10·5	22·5
2·0	4·5	3·5	2·0	8·5	3·5	32·5	2·5	20.0
1·0	2·0	5·5	4·5	7·5	14·0	8·0	23·0	3·5
7·5	13·5	17·5	5·0	31·0	13·5	16·0	22·0	15·0
6·0	25·0	60·5	5·0	25·0	12·0	43·5	2·0	8·0
13·0	12·5	8·5	13·0	3·5	3·5	3·5	2·5	12·0
7·5	12·5	0·5	3·5	19·5	1·0	8·5	2·0	40·5
16·5	4·0	17·5	34·0	14·0	13·5	5·0	7·0	67·5
3·0	0·5	22·5	7·5	10·0	13·5	1·5	4·0	0·5

Table 7.7. Intervals between successive pulses along a nerve fibre (unit 1/50 sec.). (To be read down the columns.)

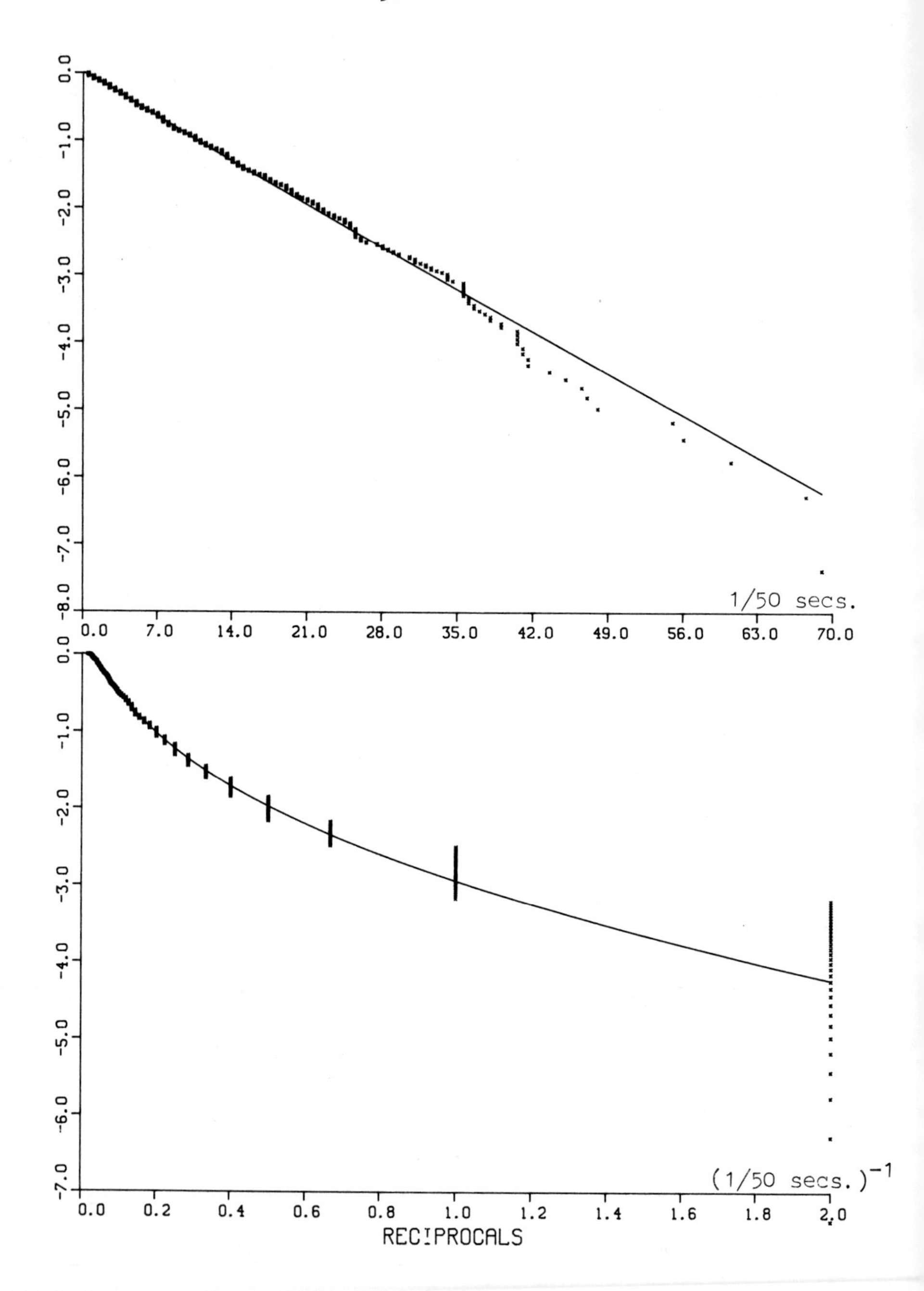

Figure 7.10. Nerve pulse data. Empirical log-survivor function and estimated log-survivor function for untransformed data (top) and reciprocal data (bottom).

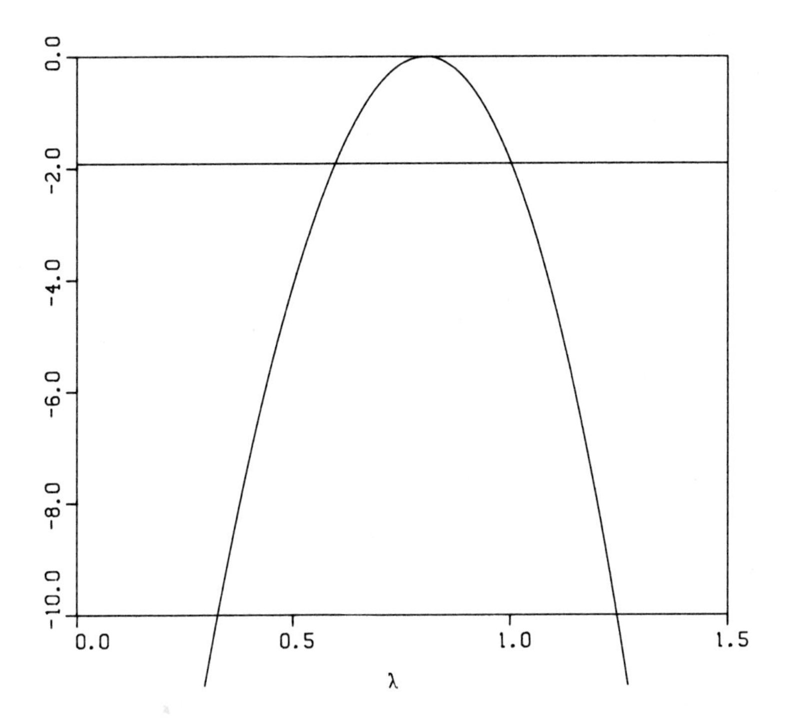

Figure 7.11. Nerve pulse data. (Normed) partially maximized log-likelihood for λ.

Figure 7.12 shows the marginal log-likelihood for ω in the case $\lambda = 1$. The likelihood is not far from being normal, and clearly indicates a significant departure from the hypothesis $\omega = 0$. The slight distortion of the log-likelihood near the maximum is caused by some numerical problems in the calculation of the Bessel function, due to the large sample size.

We have not tried to look for a normalizing transform of the parameter in this case, because the likelihood is already close to being normal.

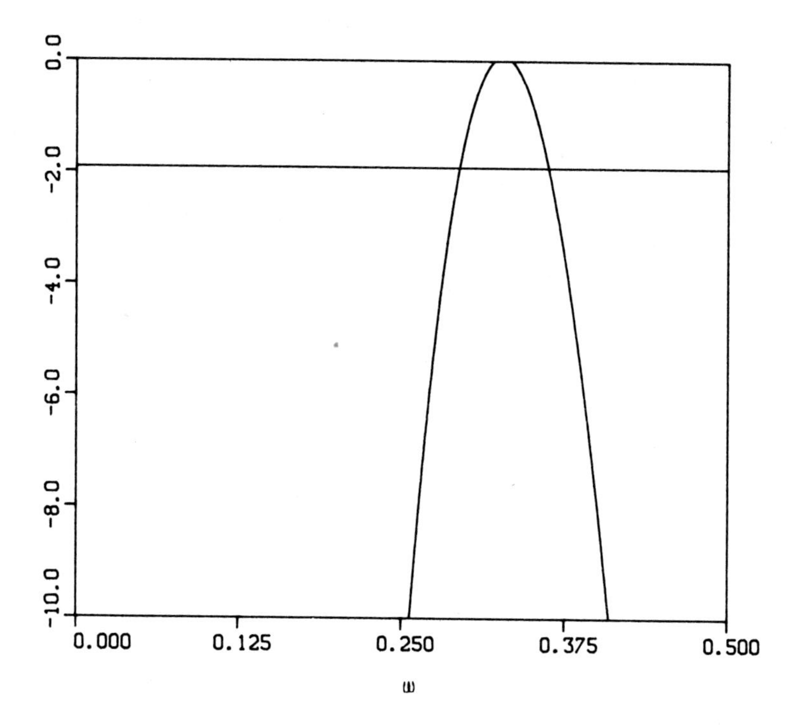

Figure 7.12. Nerve pulse data. Marginal likelihood for ω, $\lambda = 1$.

A close look at the empirical log-survivor function shows that the deviation from exponentiality occurs in the left tail of the distribution. In order to demonstrate this more clearly we have plotted the empirical log-survivor function for the reciprocal observations along with the estimated log-survivor function corresponding to an exponential distribution for the untransformed data (Figure 7.13). One observes a clear and systematic deviation in the tail of the log-survivor function, i.e. corresponding to small observations. The same discrepancy is not present in Figure 7.10, and the fit of the generalized inverse Gaussian distribution appears to be exellent. This is confirmed by the Kolmogorov-Smirnov statistic, which has the value

$$\sqrt{n}\, D_n = 0.845.$$

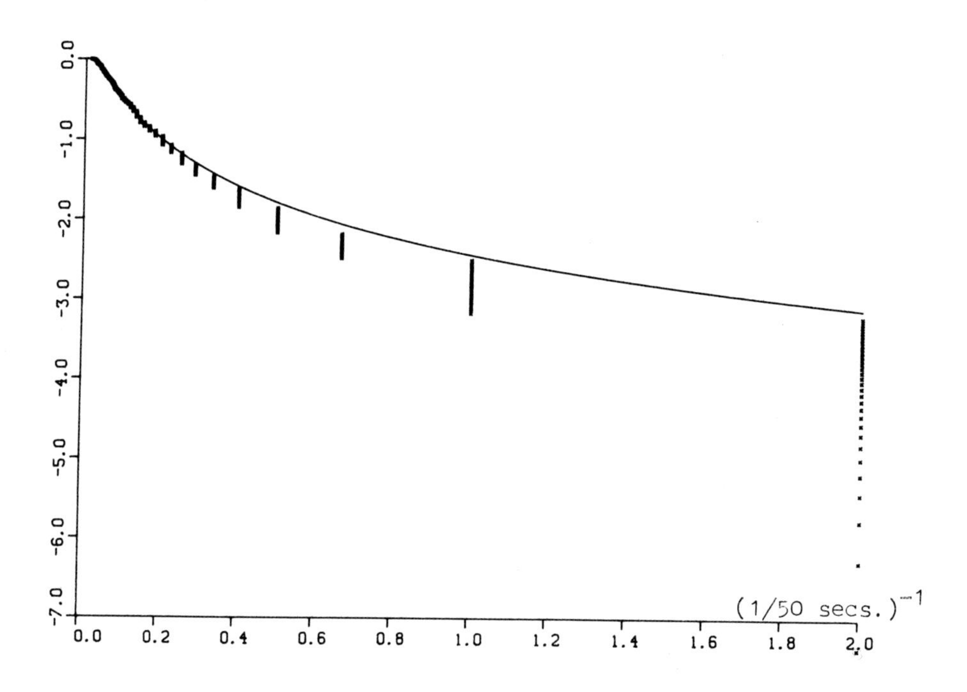

Figure 7.13. Reciprocal nerve pulse data. Empirical and estimated log-survivor function, the latter corresponding to an exponential distribution for the untransformed data.

7.3 Traffic data

Table 7.8 shows the length of 128 intervals between the times at which vehicles pass a point on a road. Figure 7.14 shows the empirical and estimated log-survivor functions, assuming a generalized inverse Gaussian renewal process, for both untransformed and reciprocal intervals.

2·8	3·4	1·4	14·5	1·9	2·8	2·3	15·3	1·8	9·5
2·5	9·4	1·1	88·6	1·6	1·9	1·5	33·7	2·6	12·9
16·2	1·9	20·3	36·8	40·1	70·5	2·0	8·0	2·1	3·2
1·7	56·5	23·7	2·4	21·4	5·1	7·9	20·1	14·9	5·6
51·7	87·1	1·2	2·7	1·0	1·5	1·3	24·7	72·6	119·8
1·2	6·9	3·9	1·6	3·0	1·8	44·8	5·0	3·9	125·3
22·8	1·9	15·9	6·0	20·6	12·9	3·9	13·0	6·9	2·5
12·3	5·7	11·3	2·5	1·6	7·6	2·3	6·1	2·1	34·7
15·4	4·6	55·7	2·2	6·0	1·8	1·9	1·8	42·0	9·3
91·7	2·4	30·6	1·2	8·8	6·6	49·8	58·1	1·9	2·9
0·5	1·2	31·0	11·9	0·8	1·2	0·8	4·7	8·3	7·3
8·8	1·8	3·1	0·8	34·1	3·0	2·6	3·7	41·3	29·7
17·6	1·9	13·8	40·2	10·1	11·9	11·0	0·2		

Table 7.8. Traffic data. The times between events are given in secs. (To be read across the rows.)

The empirical log-survivor function has the typical concave-convex shape that corresponds to a unimodal hazard, and moreover the tail appears to be linear. Since the analysis in Cox and Lewis (1966) showed no clear deviations from the renewal hypothesis it seems relevant to examine the fit of a generalized inverse Gaussian renewal model.

The partially maximized log-likelihood for λ (Figure 7.15) shows that the 95% confidence interval for λ includes the values -0.5 and 0. The maximum likelihood estimates are

$$\hat{\lambda} = -0.236, \qquad \hat{\chi} = 2.62, \qquad \hat{\psi} = 0.0267.$$

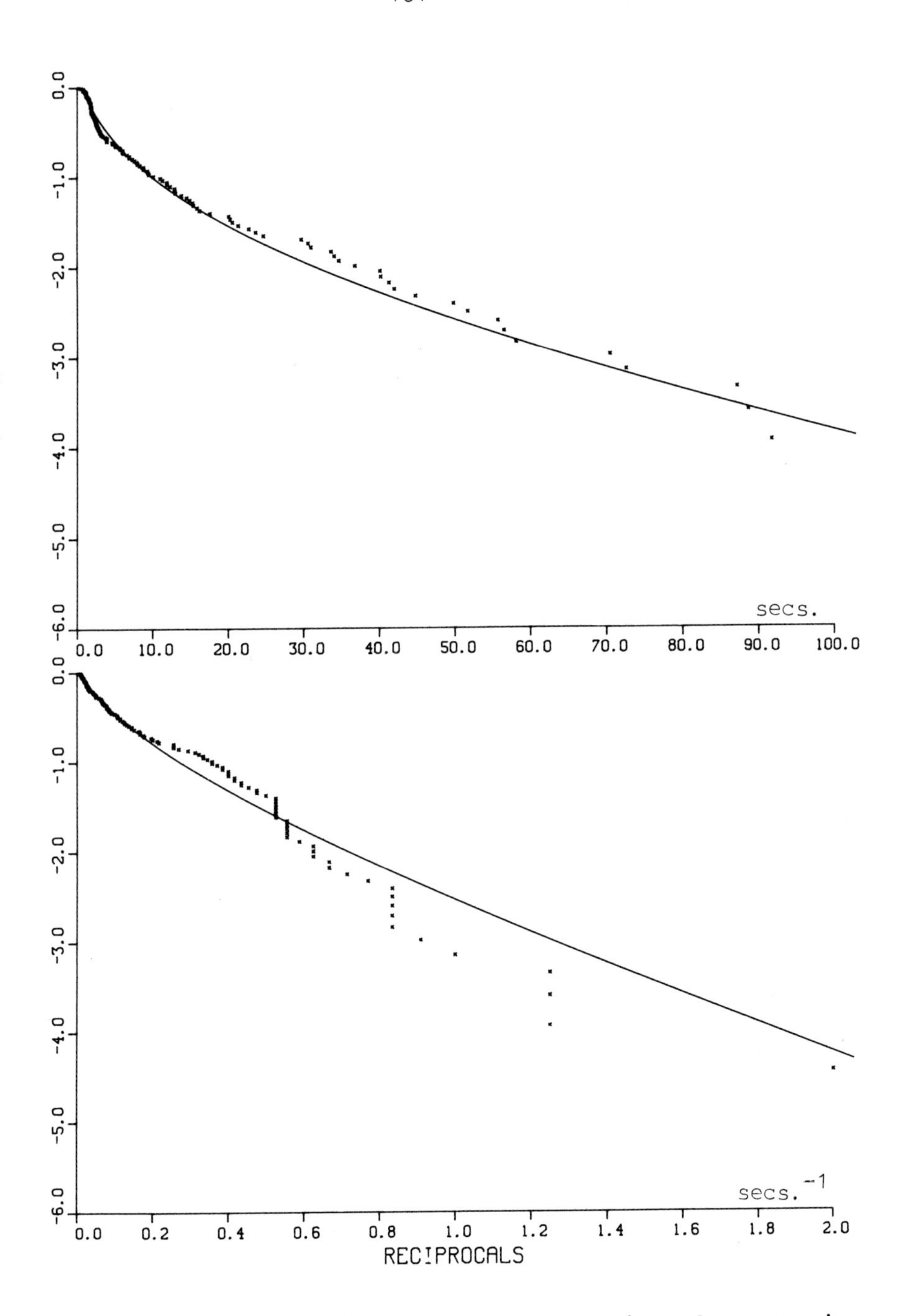

Figure 7.14. Traffic data. Empirical and estimated log-survivor functions (generalized inverse Gaussian distribution) for original data (top) and reciprocal data (bottom).

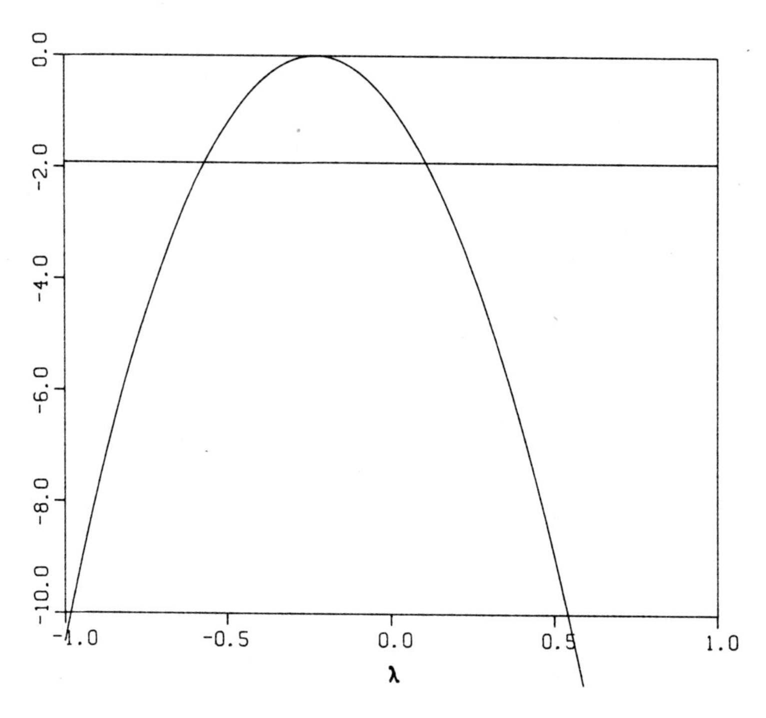

Figure 7.15. Traffic data. (Normed) partially maximized log-likelihood for λ.

Since $\lambda = 0$ is a quite plausible value for the index we have an opportunity to examine the marginal likelihood for ω in the case $\lambda = 0$. The marginal log-likelihood is shown in Figure 7.16, both as a function of ω and as a function of $\sqrt{\omega}$. It is seen that the square root transformation reduces the skewness of the likelihood and improves the normality of the likelihood. As noted earlier this might indicate the general form of a normalizing transform of ω, cf. section 7.1.

From Figure 7.14 the fit appears to be quite reasonable, except perhaps for a small interval corresponding to the abcissa 3 secs.

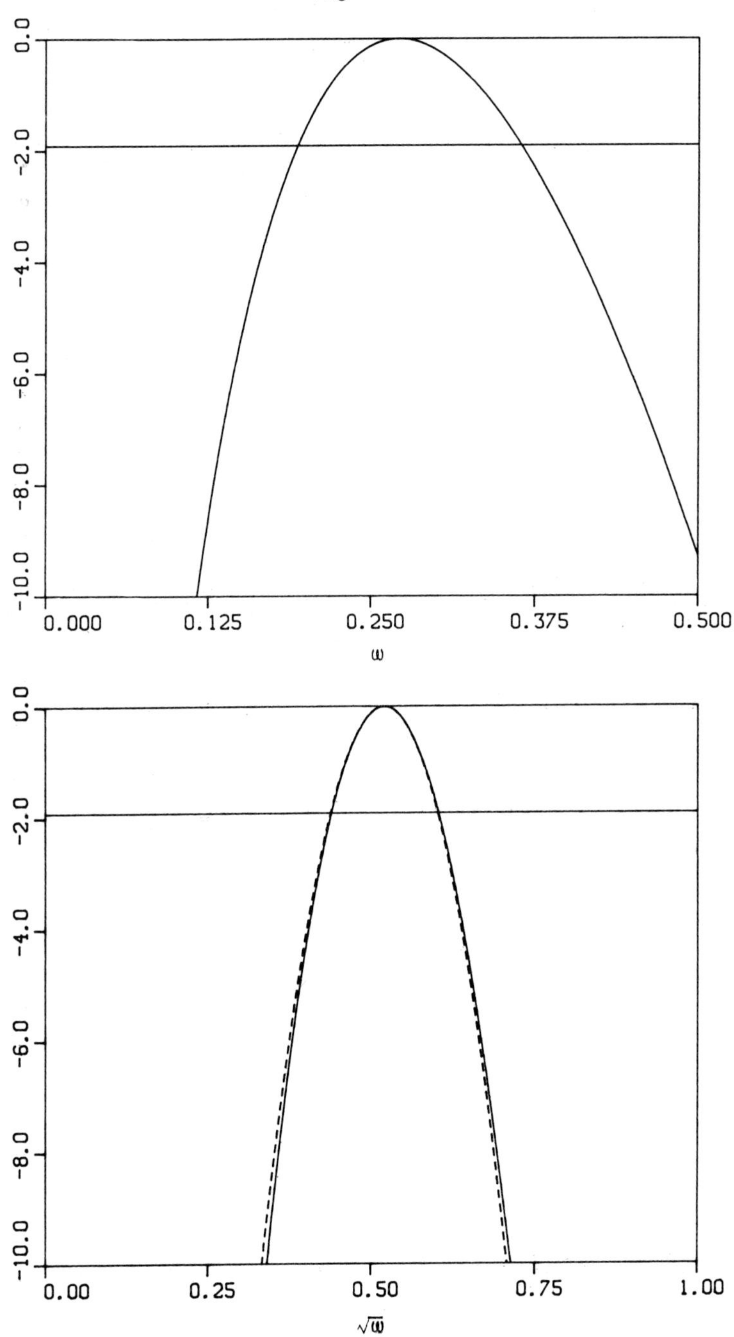

Figure 7.16. Traffic data. Marginal log-likelihood ($\lambda = 0$) for ω, as a function of ω (upper plot) and as a function of $\sqrt{\omega}$ (lower plot).

(0.33 secs.$^{-1}$ in the lower plot). However, the value of the Kolmogorov-Smirnov statistic ($\sqrt{n}\,D_n = 0.939$) does not indicate any serious discrepancies.

7.4 Repair time data

Table 7.9 shows active repair times (hours) for an airborne communication transceiver (n = 46).

.2, .3, .5, .5, .5, .5, .6, .6, .7, .7, .7, .8,
.8, 1.0, 1.0, 1.0, 1.0, 1.1, 1.3, 1.5,
1.5, 1.5, 1.5, 2.0, 2.0, 2.2, 2.5, 2.7,
3.0 3.0, 3.3, 3.3, 4.0, 4.0, 4.5, 4.7,
5.0, 5.4, 5.4, 7.0, 7.5, 8.8, 9.0, 10.3,
22.0, 24.5.

Table 7.9. Active repair times (hours) for an airborne communication transceiver.

The partially maximized log-likelihood for λ (Figure 7.17) indicated that λ should be somewhere between -1.5 and 0.5. The maximum likelihood estimate for λ ($\hat{\lambda}$ = -0.44) shows that the

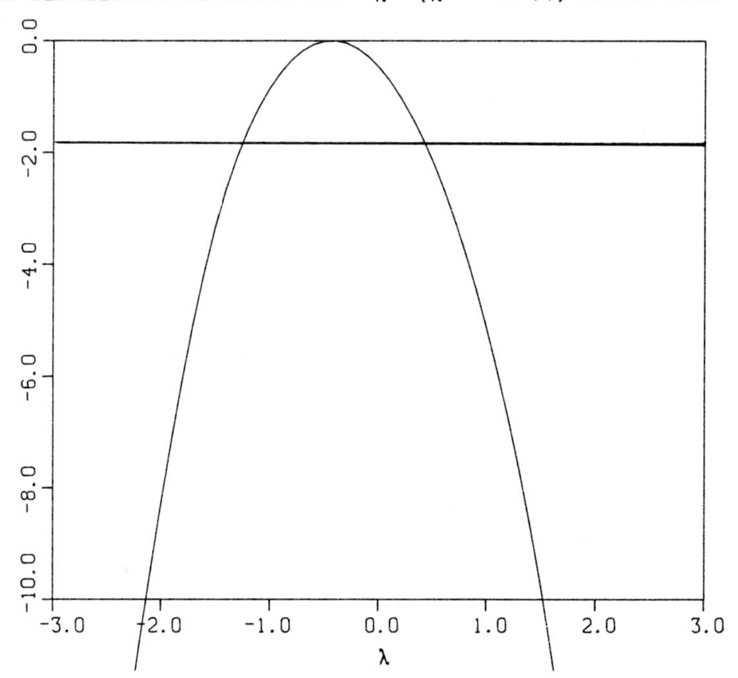

Figure 7.17. Repair time data. (Normed) partially maximized log-likelihood for λ.

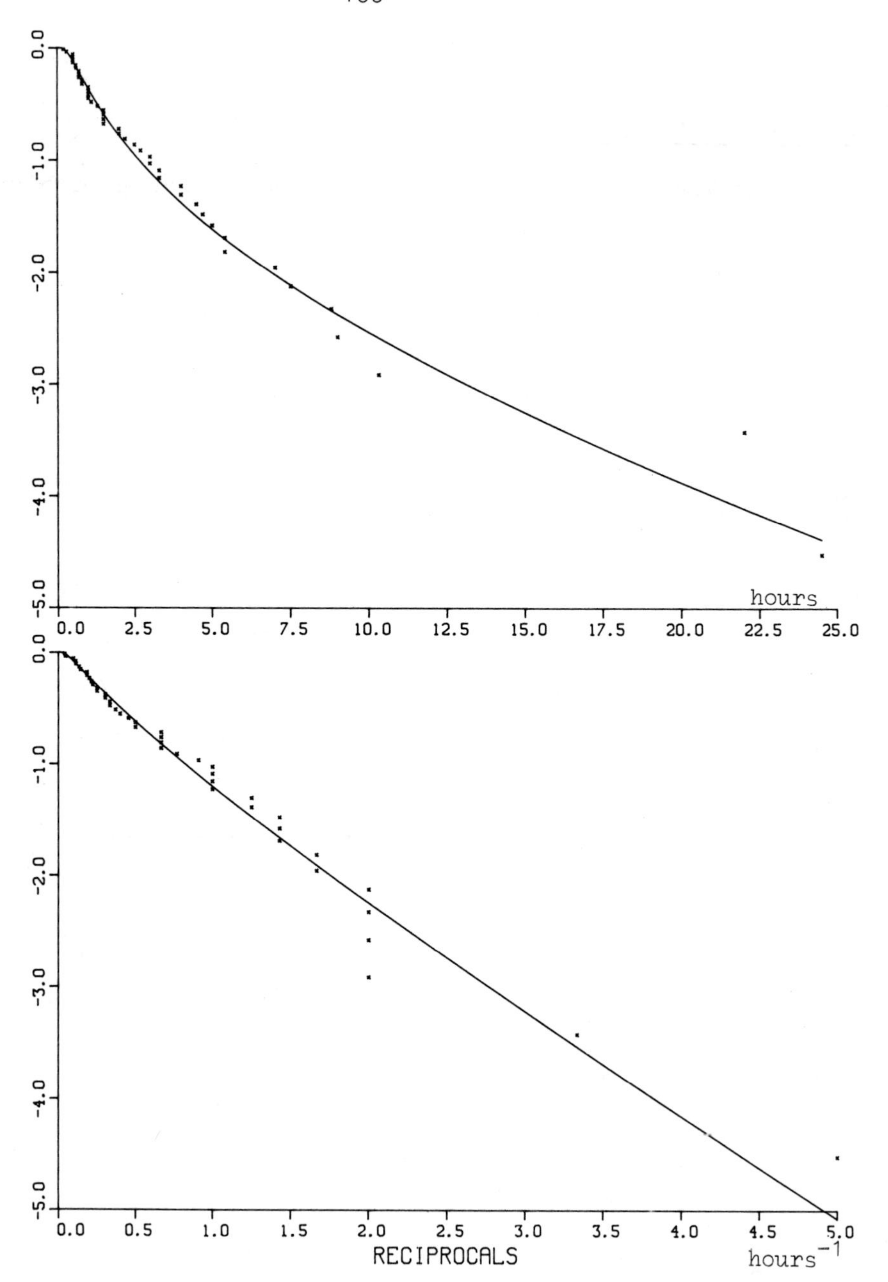

Figure 7.18. Repair time data. Empirical and estimated log-survivor function (generalized inverse Gaussian distribution) for original data (top) and reciprocal data (bottom).

distribution is not far from being inverse Gaussian, the distribution which Chhikara and Folks (1977) used in their analysis of the same data.

The plots of the empirical and estimated log-survivor functions (Figure 7.14) show that the fit is exellent, and this is confirmed by the Kolmogorov-Smirnov statistic $(\sqrt{n}\, D_n = 0.463)$. Chhikara and Folks gave a smaller value of D_n for the inverse Gaussian distribution, but we have found their calculation to be in error.

Chhikara and Folks found that the data is equally well fitted by the log normal and the inverse Gaussian distribution and it is obvious from the present example that large sample sizes are required to discriminate between for example the inverse Gaussian and the hyperbola distribution.

7.5 Fracture toughness of MIG welds

The last example concerns a sample which contains very little information about the shape of the distribution. The data gives the fracture toughness of 19 MIG welds

54.4,62.6,63.2,67.0,70.2,70.5,70.6,71.4,71.8,74.1,
74.1,74.3,78.8,81.8,83.0,84.4,85.3,86.9,87.3.

The empirical log-survivor function (Figure 7.19) shows that the observations are concentrated in a short interval. This fact is also reflected in the value of the squared mean resultant length $\bar{x}.\bar{x}_{-1} = 1.015$ being rather close to 1 ($u = 67.6$).

It is instructive to consider the partially maximized log-likelihood $\tilde{l}$ (Figure 7.20) which shows that the sample contains virtually no information about the value of λ, the partially maximized log-likelihood for λ being almost constant over the interval $-50 < \lambda < 50$. Our computer program does not work outside this interval but Figure 7.20 shows that $\hat{\lambda}$ is greater than 50. In any case such large values of λ are not of practical interest.

For large values of λ there should be little difference between the generalized inverse Gaussian distribution with $\omega > 0$ and the Gamma distribution, $\omega = 0$, (cf. section 3.1), and the appropriate test from (5.32) ($\hat{\hat{\omega}}_{50} = 44.9$, $(ni_{50}(\hat{\hat{\omega}}_{50}))^{-1/2} = 32.2$) confirms this.

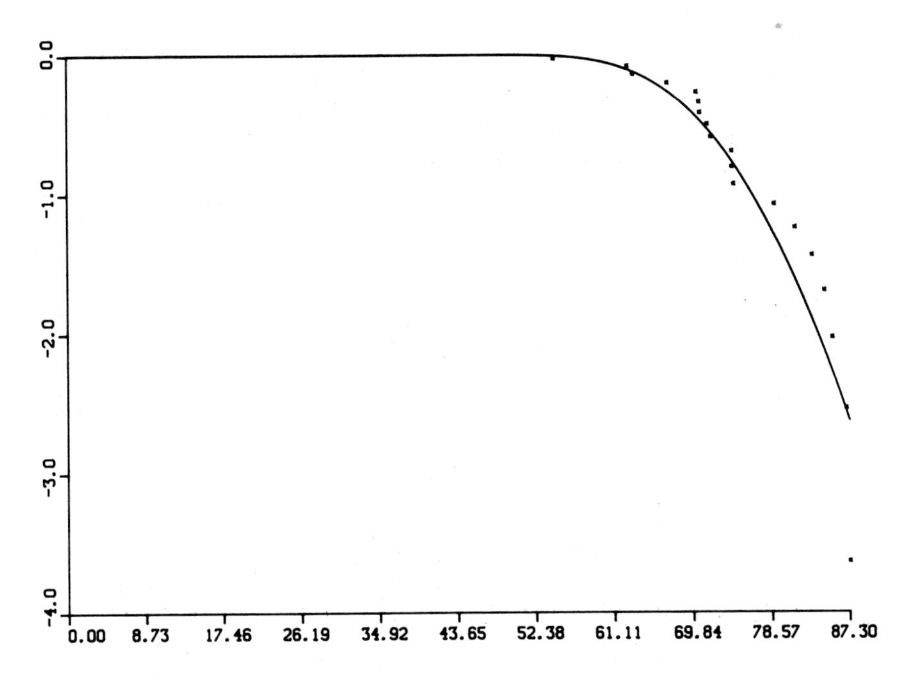

Figure 7.19. MIG welds. Empirical log-survivor function and estimated log-survivor function for $\lambda = 50$.

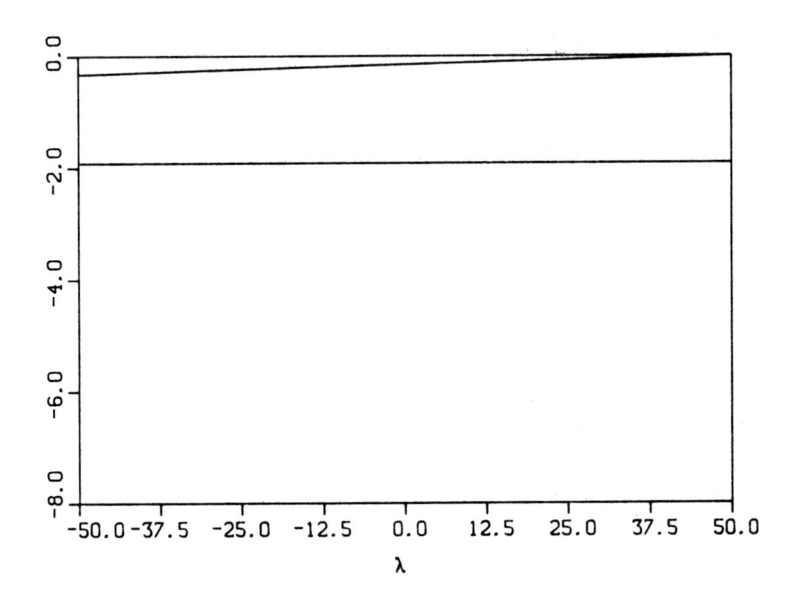

Figure 7.20. MIG welds. (Normed) partially maximized log-likelihood.

Appendix

Some results concerning the modified Bessel functions of the third kind

The modified Bessel function of the third kind and with index $\lambda \in \mathbb{R}$ is denoted by $K_\lambda(\cdot)$. Among the many integral representations of $K_\lambda(\omega)$, $\omega > 0$, one has

$$K_\lambda(\omega) = \frac{1}{2} \int_0^\infty x^{\lambda-1} e^{-\frac{1}{2}\omega(x+x^{-1})} dx.$$

In the following the results concerning K_λ, which are used in this work, will be listed. These results have mainly been extracted from Abramowitz and Stegun (1965).

The Bessel functions K_λ, $\lambda \in \mathbb{R}$, satisfy the relations

$$K_\lambda(\omega) = K_{-\lambda}(\omega) \tag{A.1}$$

$$K_{\lambda+1}(\omega) = \frac{2\lambda}{\omega} K_\lambda(\omega) + K_{\lambda-1}(\omega) \tag{A.2}$$

$$K_{\lambda-1}(\omega) + K_{\lambda+1}(\omega) = -2K_\lambda'(\omega). \tag{A.3}$$

For $\lambda = n + \frac{1}{2}$ and $n = 0,1,2,\ldots$ one has

$$K_{n+\frac{1}{2}}(\omega) = \sqrt{\frac{\pi}{2}}\,\omega^{-\frac{1}{2}} e^{-\omega}\left(1 + \sum_{i=1}^{n} \frac{(n+i)!}{(n-i)!\,i!}(2\omega)^{-i}\right). \tag{A.4}$$

The connection between K_λ and the modified Bessel function of the first kind I_ν can be expressed by

$$K_\lambda(\omega) = \frac{\pi}{2} \frac{1}{\sin(\pi\lambda)}(I_{-\lambda}(\omega) - I_\lambda(\omega)), \tag{A.5}$$

where the right hand side is to be interpreted in the limiting sense in case λ is an integer. Since

$$I_\lambda(\omega) = \sum_{m=0}^{\infty} \frac{(\frac{\omega}{2})^{2m+\lambda}}{m!\Gamma(m+\lambda+1)} \tag{A.6}$$

it is possible from (A.5) and (A.6) to derive asymptotic relations for $K_\lambda(\omega)$ as $\omega \downarrow 0$. In particular one has the first order approximation

$$K_\lambda(\omega) \simeq \Gamma(\lambda)2^{\lambda-1}\omega^{-\lambda} \quad (\omega \downarrow 0) \quad (\lambda > 0). \tag{A.7}$$

For $\lambda = 0$ one has

$$K_0(\omega) \simeq -\ln\omega \quad (\omega \downarrow 0). \tag{A.8}$$

For large ω the following asymptotic expansion of $K_\lambda(\omega)$ is valid

$$K_\lambda(\omega) = \sqrt{\tfrac{\pi}{2}}\omega^{-\frac{1}{2}}e^{-\omega}\left(1 + \frac{\mu-1}{8\omega} + \frac{(\mu-1)(\mu-9)}{2!(8\omega)^2} + \frac{(\mu-1)(\mu-9)(\mu-25)}{3!(8\omega)^3} + \ldots\right), \tag{A.9}$$

where $\mu = 4\lambda^2$.

From Ismail (1977) we have the following asymptotic relation for large λ

$$K_\lambda(\omega) \simeq 2^\lambda \lambda^{\lambda-\frac{1}{2}}e^{-\lambda}\omega^{-\lambda}\sqrt{\tfrac{\pi}{2}}, \quad (\lambda \to \infty) \tag{A.10}$$

which may be viewed as an analogue of Stirling's formula for the gamma function.

The functions D_λ and R_λ.

We define the functions R_λ and D_λ by

$$R_\lambda(\omega) = \frac{K_{\lambda+1}(\omega)}{K_\lambda(\omega)} \tag{A.11}$$

and

$$D_\lambda(\omega) = \frac{K_{\lambda+1}(\omega)K_{\lambda-1}(\omega)}{K_\lambda(\omega)^2}, \tag{A.12}$$

respectively.

The following relations are easily derived from (A.1) and (A.2)

$$D_\lambda(\omega) = D_{-\lambda}(\omega) \tag{A.13}$$

$$D_\lambda(\omega) = R_\lambda(\omega)R_{-\lambda}(\omega) \tag{A.14}$$

$$R_{-\lambda}(\omega) = R_{\lambda-1}(\omega)^{-1} \tag{A.15}$$

$$R_\lambda(\omega) = \frac{2\lambda}{\omega} + R_{-\lambda}(\omega) \tag{A.16}$$

$$R_\lambda(\omega) = \frac{\lambda}{\omega} + \sqrt{(\tfrac{\lambda}{\omega})^2 + D_\lambda(\omega)}\ . \tag{A.17}$$

If $\lambda + 1/2$ is an integer it follows from (A.4) that R_λ and D_λ are rational functions, and the most simple cases of R_λ and D_λ are

$$\begin{aligned} R_{-1/2}(\omega) &= 1 \\ R_{1/2}(\omega) &= 1 + \frac{1}{\omega} \\ R_{-3/2}(\omega) &= \frac{\omega}{\omega+1} \\ D_{\pm 1/2}(\omega) &= 1 + \frac{1}{\omega}\ . \end{aligned} \tag{A.18}$$

It was shown by Lorch (1967) that the function

$$\frac{K_{\lambda+\epsilon}(\cdot)}{K_\lambda(\cdot)}$$

is decreasing if $\lambda \geq 0$ and $\epsilon > 0$ and it follows that $R_\lambda(\cdot)$ is decreasing if $\lambda > -\frac{1}{2}$ and increasing if $\lambda < -\frac{1}{2}$. Ismail and Muldoon (1978) have proved that $R_\lambda(\omega)$ is an increasing function of λ for given $\omega > 0$. In Theorem 4.1 it is shown that $D_\lambda(\cdot)$ is decreasing for any given λ.

By (A.7) and (A.9) we have

$$\lim_{\omega \downarrow 0} R_\lambda(\omega) = \begin{cases} \infty & \text{if} \quad \lambda > -1/2 \\ 0 & \text{if} \quad \lambda < -1/2 \end{cases} \tag{A.19}$$

and

$$\lim_{\omega \downarrow 0} D_\lambda(\omega) = \begin{cases} \infty & \text{if} \quad |\lambda| \leq 1 \\ \frac{|\lambda|}{|\lambda|-1} & \text{if} \quad |\lambda| > 1 . \end{cases} \tag{A.20}$$

The following asymptotic expansions can be derived from (A.9)

$$R_\lambda(\omega) = 1 + \frac{8\lambda+4}{8\omega} + \frac{32\lambda^2-8}{(8\omega)^2} + \frac{-256\lambda^2+64}{(8\omega)^3} + O(\omega^{-4}) \qquad (\omega \to \infty) \tag{A.21}$$

$$D_\lambda(\omega) = 1 + \frac{1}{\omega} + \frac{-256\lambda^2+64}{(8\omega)^3} + O(\omega^{-4}) \qquad (\omega \to \infty). \tag{A.22}$$

From (A.3) and (A.2) the following expressions for the derivatives of R_λ and D_λ can be found

$$R_\lambda'(\omega) = R_\lambda^2(\omega) - \frac{2\lambda+1}{\omega} R_\lambda(\omega) - 1 \tag{A.23}$$

$$D_\lambda'(\omega) = D_\lambda(\omega)(R_\lambda(\omega) - R_\lambda(\omega)^{-1} + R_{-\lambda}(\omega) - R_{-\lambda}(\omega)^{-1} - \frac{2}{\omega}). \tag{A.24}$$

Plots of R_λ and D_λ are shown in Figure A.1 and A.2, respectively, for some values of λ.

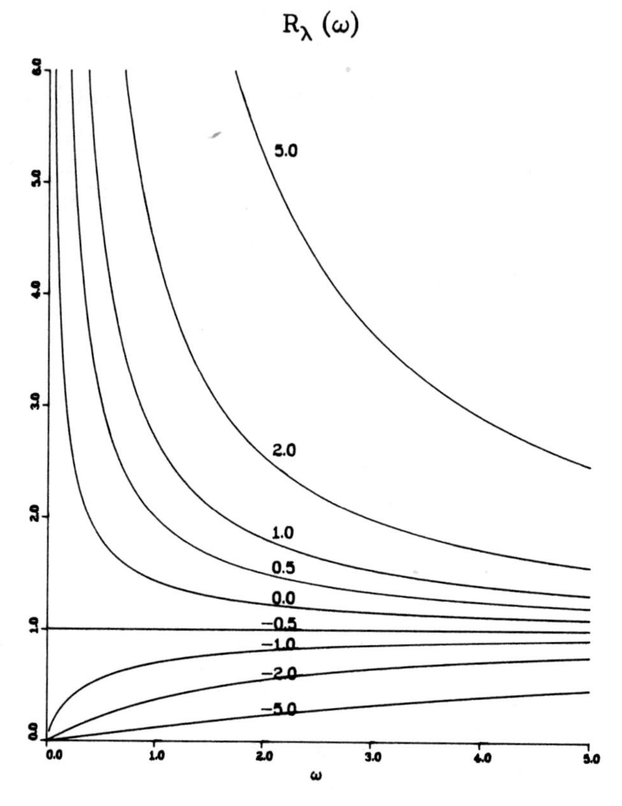

Figure A.1. Plots of $R_{\lambda}(\cdot)$ (the value of λ is indicated at each curve).

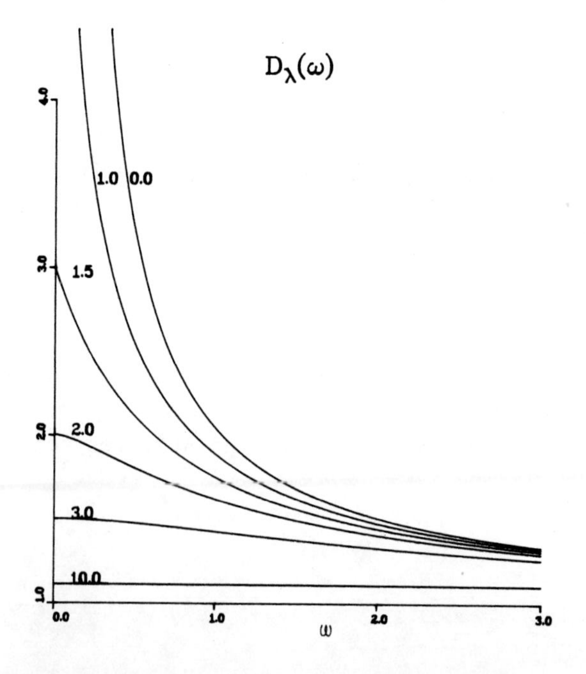

Figure A.2. Plots of $D_{\lambda}(\cdot)$ (the value of λ is indicated at each curve).

The following theorem has been proved by P.W.Karlsson, Technical University of Denmark (personal communication).

Theorem A.1. For a fixed value of $x \in \mathbb{R}_+$ the function $D_\lambda(x)$ is increasing as a function of λ for $\lambda < 0$ and decreasing as a function of λ for $\lambda > 0$. **

Proof. Introduce the function

$$Z_\lambda(x) = \frac{K'_\lambda(x)}{K_\lambda(x)} = \frac{\frac{\lambda}{x} K_\lambda(x) - K_{\lambda+1}(x)}{K_\lambda(x)} = \frac{\lambda}{x} - R_\lambda(x),$$

and note that from (A.16)

$$R_{-\lambda}(x) = -\frac{2\lambda}{x} + R_\lambda(x).$$

Hence (A.12) may be rewritten as follows

$$D_\lambda(x) = R_\lambda(x)(R_\lambda(x) - \frac{2\lambda}{x}) = (\frac{\lambda}{x} - Z_\lambda(x))(-\frac{\lambda}{x} - Z_\lambda(x)),$$

or

$$D_\lambda(x) = Z_\lambda(x)^2 - \frac{\lambda^2}{x^2}. \tag{A.25}$$

An integral representation for $Z_\lambda(x)^2$ will now be derived.

The functions K_λ and I_λ are linearly independent solutions of the modified Bessel equation

$$y'' + \frac{1}{x}y' - (1 + \frac{\lambda^2}{x^2})y = 0. \tag{A.26}$$

Following Amos (1974, p.240) we now obtain, for a solution to (A.26)

$$\frac{1}{2}\frac{d}{dx}[(xy')^2] = (xy')(xy')' = x^2y'y'' + xy'^2 = (x^2+\lambda^2)yy',$$

i.e.

$$\frac{d}{dx}[(xy')^2] = (x^2+\lambda^2)\frac{d}{dx}y^2. \tag{A.27}$$

Since K_λ and K'_λ for λ fixed vanish exponentially as $x \to +\infty$, equation (A.27) after a partial integration leads to

$$[\tilde{x}^2 K'_\lambda(\tilde{x})^2]_x^{+\infty} = [(\tilde{x}^2+\lambda^2)K_\lambda(\tilde{x})^2]_x^{+\infty} - 2\int_x^{+\infty} \tilde{x} K_\lambda(\tilde{x})^2 d\tilde{x}$$

and, after division,

$$Z_\lambda(x)^2 = 1 + \frac{\lambda^2}{x^2} + \frac{2}{x^2 K_\lambda(x)^2}\int_x^{+\infty} \tilde{x} K_\lambda(\tilde{x})^2 d\tilde{x} \quad (x \in \mathbb{R}_+), \tag{A.28}$$

this is the analogue of equation (6) of Amos (1974). Comparison with (A.25) yields

$$D_\lambda(x) = 1 + \frac{2}{x^2}\int_x^{+\infty} \tilde{x}\left(\frac{K_\lambda(\tilde{x})}{K_\lambda(x)}\right)^2 d\tilde{x} \quad (x \in \mathbb{R}_+). \tag{A.29}$$

It has been shown by Hartman (1976) that the quotient $K_\lambda(\tilde{x})/K_\lambda(x)$ is a completely monotonic function of λ^2 for $0<x<\tilde{x}<\infty$. In particular, $K_\lambda(\tilde{x})/K_\lambda(x)$ is a decreasing function of λ^2, and it follows from (A.29) that $D_\lambda(x)$ is decreasing (increasing) for λ positive (negative). **

References

The list includes a number of titles not cited in the text, particularly on the inverse Gaussian distribution.

Abramowitz, M. and Stegun, I.A. (1965): Handbook of Mathematical Functions. Dover Publications, New York.

Amos, D.E. (1974): Computation of modified Bessel functions and their ratios. Math.Comp. 28, 239-251.

Atkinson, A.C. (1979): The simulation of generalized inverse Gaussian, generalized hyperbolic, gamma and related random variables. Research Report No.52, Dept.Theor.Statist., Aarhus University.

Banerjee, A.K. and Bhattacharyya, G.K. (1979): Baysian results for the inverse Gaussian distribution with an application. Technometrics 21, 247-252.

Barndorff-Nielsen, O. (1977): Exponentially decreasing log-size distributions. Proc.R.Soc.London A 353, 401-419.

Barndorff-Nielsen, O. (1978a): Information and exponential families in statistical theory. Wiley, Chichester.

Barndorff-Nielsen, O. (1978b): Hyperbolic distributions and distributions on hyperbolae. Scand.J.Statist. 5, 151-157.

Barndorff-Nielsen, O., Blæsild, P. and Halgreen, C. (1978): First hitting time models for the generalized inverse Gaussian distribution. Stoch.Processes Appl. 7, 49-54.

Barndorff-Nielsen, O. and Cox, D.R. (1979): Edgeworth and saddle-point approximations with statistical applications. (With discussion.) J.R.Statist.Soc. B 41, 279-312.

Barndorff-Nielsen, O. and Halgreen, C. (1977): Infinite divisibility of the hyperbolic and generalized inverse Gaussian distributions. Z.Wahrscheinlichkeitstheorie Verw.Gebiete 38, 309-311.

Blæsild, P. (1978): The shape of the generalized inverse Gaussian and hyperbolic distributions. Research Report No.37, Dept. Theor.Statist., Aarhus University.

Box, G.E.P. and Tiao, G.C. (1973): Bayesian inference in statistical analysis. Addison-Wesley, Reading.

Chhikara, R.S. and Folks, J.L. (1977): The inverse Gaussian distribution as a lifetime model. Technometrics 19, 461-468.

Chhikara, R.S. and Folks, J.L. (1978): The inverse Gaussian distribution and its statistical application - a review. (With discussion). J.R.Statist.Soc. B 40, 263-289.

Cox, D.R. and Lewis, P.A.W. (1966): The statistical analysis of series of events. Methuen, London.

Davis, H.T. and Feldstein, M.L. (1979): The generalized Pareto law as a model for progressively censored survival data. Biometrika 66, 299-306.

Eaton, W.W. and Whitmore, G.A. (1977): Length of stay as a stochastic process: A general approach and application to hospitalization for Schizophrenia. Journal of Mathematical Sociology 5, 273-292.

Efron, B. and Hinkley, D.V. (1978): Assessing the accuracy of the maximum likelihood estimator: Observed versus expected Fisher information. Biometrika 65, 457-487.

Faxén, H. (1920): Expansion in series of the integral $\int_y^\infty e^{-x(t\pm t^{-\mu})} t^\nu dt$. Arkiv för matematik, astronomi och fysik 15, 1-57.

Fisher, R.A. (1956): Statistical methods and scientific inference. Oliver and Boyd, Edinburgh.

Good, I.J. (1953): The population frequencies of species and the estimation of population parameters. Biometrika 40, 237-60.

Hadwiger, H. (1940): Eine analytische Reproduktionsfunktion für biologische Gesamtheiten. Skandinavisk Aktuarietidskrift 23, 101-113.

Halgreen, C. (1979): Self-decomposability of the generalized inverse Gaussian and hyperbolic distributions. Z.Wahrscheinlichkeitstheorie Verw.Gebiete 47, 13-18.

Hartman, P. (1976): Completely monotone families of solutions of n-th order linear differential equations and infinitely divisible distributions. An.Scuola Norm.Sup.Pisa Cl.Sci. (4) 3, 267-287.

Hoem, J.M. (1976): The statistical theory of demographic rates. A review of current developments. (With discussion.) Scand. J.Statist. 3, 169-185.

Ismail, M.E.H. (1977): Integral representations and complete monotonicity of various quotients of Bessel functions. Can.J.Math. 24, 1198-1207.

Ismail, M.E.H. and Muldoon, M.E. (1978): Monotonicity of the zeros of a cross-product of Bessel functions Siam.J.Math.Anal. 9, 759-767.

Johnson, N.L. and Kotz, S. (1970): Distributions in statistics: continuous univariate distributions 1. Boston: Houghton-Mifflin.

Jørgensen, B. and Pedersen, B.V. (1979): Contribution to the discussion of O.Barndorff-Nielsen and D.R.Cox: Edgeworth and saddle-point approximations with statistical applications. J.R.Statist.Soc. B 41, 309-310.

Keyfitz, N. (1968): Introduction to the mathematics of population. Addison-Wesley, Reading.

Lancaster, T. (1972): A stochastic model for the duration of a strike. J.R.Statist.Soc. A 135, 257-271.

Lombard, F. (1978): A sequential test for the mean of an inverse Gaussian distribution. South African Statist. J. 12, 107-115.

Lorch, L. (1967): Inequalities for some Whittaker functions. Arch. Math. (Brno) 3, 1-9.

Marcus, A.H. (1975): Power laws in compartmental analysis. Part I: a unified stochastic model. Math.Biosci. 23, 337-350.

Mardia, K.V. (1972): Statistics of directional data. London: Academic Press.

Miura, C.K. (1978): Tests for the mean of the inverse Gaussian distribution. Scand.J.Statist. 5, 200-204.

Nádas, A. (1973): Best tests for zero drift based on first passage times in Brownian motion. Technometrics 15, 125-132.

Olver, F.W.J. (1974): Asymptotics and special functions. Academic Press, New York.

Padgett, W.J. and Wei, L.J. (1979): Estimation for the three-parameter inverse Gaussian distribution. Commun.Statist.-Theor.Meth. A8, 129-137.

Patil, S.A. and Kovner, J.L. (1979): On the pover of an optimum test for the mean of the inverse Gaussian distribution. Technometrics 21, 379-381.

Rukhin, A.L. (1974): Strongly symmetric families and statistical analysis of their parameters. Zap.Naucvn.Sem.Leningrad.Otdel. Mat.Inst.Steklov 43, 59-87. (English translation (1978): J.Soviet Math. 9, 886-910.)

Schou, G. (1978): Estimation of the concentration parameter in the von Mises-Fisher distributions. Biometrika 65, 369-377.

Schrödinger, E. (1915): Zur Theorie der Fall - und Steigversuche an Teilschen mit Brownscher Bewegung. Physikalische Zeitschrift 16, 289-295.

Seshadri, V. and Shuster, J.J. (1974): Exact tests for zero drift based on first passage times in Brownian motion. Technometrics 16, 133-134.

Shuster, J.J. (1968): On the inverse Gaussian distribution function. J.Amer.Statist.Ass. 63, 1514-1516.

Sichel, H.S. (1974): On a distribution representing sentence-length in written prose. J.R.Statist.Soc. A 137, 25-34.

Sichel, H.S. (1975): On a distribution law for word frequencies. J.Amer.Statist.Ass. 70, 542-547.

Sprott, D.A. (1973): Normal likelihoods and their relation to large sample theory of estimation. Biometrika 60, 457-465.

Tweedie, M.C.K. (1957): Statistical properties of inverse Gaussian distributions I. Ann.Math.Statist. 28, 362-377.

Whitmore, G.A. and Yalovsky, M. (1978): A normalizing logarithmic transformation for inverse Gaussian random variables. Technometrics 20, 207-208.

Whitmore, G.A.(1979): An inverse Gaussian model for labour turn-over. J.R.Statist.Soc. A 142, 468-478.

Wise, M.G. (1975): Skew distributions in biomedicine including some with negative powers of time. In Statistical Distributions in Scientific Work, Vol.2: Model building and Model selection (G.P.Patil et al., eds.), 241-262. Dordrect Reidel.

Subject index

List of symbols

Symbol	Description	Page
χ, ψ	parameters	1
λ	index parameter	1
ω	concentration parameter	5
η	scale parameter	5
Θ_λ	parameter domain for (χ, ψ)	5
$N^{-}(\lambda, \chi, \psi)$	generalized inverse Gaussian distribution (1.1)	5
N_λ^{-}	the class of generalized inverse Gaussian distributions with index λ	5
$(X_\sim, X_{-}, X.) = (\sum_{i=1}^{n} \ln X_i, \sum_{i=1}^{n} X_i^{-1}, \sum_{i=1}^{n} X_i)$		11
$S = \sqrt{X./X_{-}}$	direction	67
$T = \sqrt{X.X_{-}}$	resultant length	67
$W = \bar{X}_\sim - \ln S$		77
$\hat{\chi}_\lambda, \hat{\psi}_\lambda, \hat{\omega}_\lambda, \hat{\eta}_\lambda$	maximum likelihood estimates, fixed λ	41
$(\hat{\lambda}, \hat{\chi}, \hat{\psi})$	maximum likelihood estimate	58
$i_\lambda(\chi, \psi)$	Fisher information about (χ, ψ), fixed λ	51
$i(\lambda, \chi, \psi)$	Fisher information about (λ, χ, ψ)	62
$\tilde{l}(\lambda)$	partially maximized log-likelihood for λ	58

$u = \bar{x}.\bar{x}_{\dashv}/(\bar{x}.\bar{x}_{\dashv} - 1)$ 42

$\hat{\hat{\omega}}_\lambda$ marginal maximum likelihood estimate for ω 85

$i_\lambda(\omega)$ expected information for ω 86

$\hat{j}_\lambda$ observed information for ω 84

$l_\lambda(\omega)$ marginal log-likelihood for ω 83

$|F_\omega|$ Sprott's measure for the deviation of l_λ from normality 84

$K_\lambda(\omega)$ modified Bessel function of the third kind 170

$D_\lambda(\omega)$ 172

$R_\lambda(\omega)$ 171

Lecture Notes in Statistics

Vol. 1: R. A. Fisher: An Appreciation. Edited by S. E. Fienberg and D. V. Hinkley. xi, 208 pages, 1980.

Vol. 2: Mathematical Statistics and Probability Theory. Proceedings 1978. Edited by W. Klonecki, A. Kozek, and J. Rosiński. xxiv, 373 pages, 1980.

Vol. 3: B. D. Spencer, Benefit-Cost Analysis of Data Used to Allocate Funds. viii, 296 pages, 1980.

Vol. 4: E. A. van Doorn: Stochastic Monotonicity and Queueing Applications of Birth-Death Processes. vi, 118 pages, 1981.

Vol. 5: T. Rolski, Stationary Random Processes Associated with Point Processes. vi, 139 pages, 1981.

Vol. 6: S. S. Gupta and D.-Y. Huang, Multiple Statistical Decision Theory: Recent Developments. viii, 104 pages, 1981.

Vol. 7: M. Akahira and K. Takeuchi, Asymptotic Efficiency of Statistical Estimators. viii, 242 pages, 1981.

Vol. 8: The First Pannonian Symposium on Mathematical Statistics. Edited by P. Révész, L. Schmetterer, and V. M. Zolotarev. vi, 308 pages, 1981.

Vol. 9: B. Jørgensen, Statistical Properties of the Generalized Inverse Gaussian Distribution. vi, 188 pages, 1981.

Springer Series in Statistics

L. A. Goodman and W. H. Kruskal, Measures of Association for Cross Classifications. x, 146 pages, 1979.

J. O. Berger, Statistical Decision Theory: Foundations, Concepts, and Methods. xiv, 420 pages, 1980.

R. G. Miller, Jr., Simultaneous Statistical Inference, 2nd edition. xvi, 299 pages, 1981.

P. Brémaud, Point Processes and Queues: Martingale Dynamics. xviii, 354 pages, 1981.

E. Seneta, Non-Negative Matrices and Markov Chains. xv, 279 pages, 1981.

F. J. Anscombe, Statistical Computing with APL. xvi, 426 pages, 1981.

J. W. Pratt and J. D. Gibbons, Concepts of Nonparametric Theory. xvi, 462 pages, 1981.

Lecture Notes in Mathematics

Selected volumes of interest to statisticians and probabilists:

Vol. 532: Théorie Ergodique. Actes des Journées Ergodiques, Rennes 1973/1974. Edited by J.-P. Conze, M. S. Keane. 227 pages, 1976.

Vol. 539: Ecole d'Ete de Probabilités de Saint-Flour V—1975. A. Badrikian, J. F. C. Kingman, J. Kuelbs. Edited by P.-L Hennequin. 314 pages, 1976.

Vol. 550: Proceedings of the Third Japan-USSR Symposium on Probability Theory. Edited by G. Maruyama, J. V. Prokhorov. 722 pages, 1976.

Vol. 566: Empirical Distributions and Processes. Selected Papers from a Meeting at Oberwolfach, March 28—April 3, 1976. Edited by P. Gänssler, P. Revesz. 146 pages, 1976.

Vol. 581: Séminaire de Probabilités XI. Université de Strasbourg. Edited by C. Dellacherie, P. A. Meyer, and M. Weil. 573 pages, 1977.

Vol. 595: W. Hazod, Stetige Faltungshalbgruppen von Warscheinlichkeitsmassen und erzeugende Distributionen. 157 pages, 1977.

Vol. 598: Ecole d'Ete de Probabilités de Saint-Flour VI—1976. J. Hoffmann-Jorgenser, T. M. Ligett, J. Neveu. Edited by P.-L Hennequin. 447 pages, 1977.

Vol. 636: Journées de Statistique des Processus Stochastiques, Grenoble 1977. Edited by Didier Dacunha-Castelle, Bernard Van Cutsem. 202 pages, 1978.

Vol. 649: Séminaire de Probabilités XII. Strasbourg 1976–1977. Edited by C. Dellacherie, P. A. Meyer, and M. Weil. 805 pages, 1978.

Vol. 656: Probability Theory on Vector Spaces, Proceedings 1977. Edited by A. Weron. 274 pages. 1978.

Vol. 672: R. L. Taylor, Stochastic Convergence of Weighted Sums of Random Elements in Linear Spaces. 216 pages, 1978.

Vol. 675: J. Galambos, S. Kotz, Characterizations of Probability Distributions. 169 pages, 1978.

Vol. 678: Ecole d'Ete de Probabilités de Saint-Flour VII—1977. D. Dacunha-Castelle, H. Heyer, and B. Roynette. Edited by P.-L. Hennequin. 379 pages, 1978.

Vol. 690: W. J. J. Rey, Robust Statistical Methods. 128 pages, 1978.

Vol. 706: Probability Measures on Groups, Proceedings 1978. Edited by H. Heyer. 348 pages, 1979.

Vol. 714: J. Jacod, Calcul Stochastique et Problemes de Martingales. 539 pages, 1979.

Vol. 721: Séminaire de Probabilités XIII. Proceedings, Strasbourg, 1977/78. Edited by C. Dellacherie, P. A. Meyer, and M. Weil. 647 pages, 1979.

Vol. 794: Measure Theory, Oberwolfach 1979, Proceedings, 1979. Edited by D. Kolzöw. 573 pages, 1980.

Vol. 796: C. Constantinescu, Duality in Measure Theory. 197 pages, 1980.

Vol. 821: Statistique non Paramétrique Asymptotique, Proceedings 1979. Edited by J.-P. Raoult. 175 pages, 1980.

Vol. 828: Probability Theory on Vector Spaces II, Proceedings 1979. Edited by A. Weron. 324 pages, 1980.